Towards Human Brain Inspired Lifelong Learning

Towards Human Brain Inspired Lifelong Learning

editors

Xiaoli Li
*A*STAR, Singapore*

Savitha Ramasamy
*A*STAR, Singapore*

ArulMurugan Ambikapathi
Lam Research, Singapore

Suresh Sundaram
Indian Institute of Science, Bangalore, India

Haytham M. Fayek
RMIT University, Australia

NEW JERSEY · LONDON · SINGAPORE · BEIJING · SHANGHAI · HONG KONG · TAIPEI · CHENNAI · TOKYO

Published by

World Scientific Publishing Co. Pte. Ltd.
5 Toh Tuck Link, Singapore 596224
USA office: 27 Warren Street, Suite 401-402, Hackensack, NJ 07601
UK office: 57 Shelton Street, Covent Garden, London WC2H 9HE

Library of Congress Cataloging-in-Publication Data
Names: Li, Xiaoli, 1970–   editor. | Ramasamy, Savitha editor. | Ambikapathi, ArulMurugan,
  editor. | Sundaram, Suresh (Aerospace engineer), editor. | Fayek, Haytham M., editor.
Title: Towards human brain inspired lifelong learning / Xiaoli Li (A*STAR, Singapore),
  Savitha Ramasamy (A*STAR, Singapore), ArulMurugan Ambikapathi
  (Lam Research, Singapore), Suresh Sundaram (Indian Institute of Science, Bangalore, India),
  Haytham M. Fayek (RMIT University, Australia).
Description: New Jersey : World Scientific, [2024] | Includes bibliographical references and index.
Identifiers: LCCN 2023057768 | ISBN 9789811286704 (hardcover) |
  ISBN 9789811286711 (ebook for institutions) | ISBN 9789811286728 (ebook for individuals)
Subjects: LCSH: Machine learning. | Neural networks (Computer science) |
  Deep learning (Machine learning)
Classification: LCC Q325.5 .T693 2024 | DDC 006.3/1--dc23/eng/20240201
LC record available at https://lccn.loc.gov/2023057768

British Library Cataloguing-in-Publication Data
A catalogue record for this book is available from the British Library.

Copyright © 2024 by World Scientific Publishing Co. Pte. Ltd.

All rights reserved. This book, or parts thereof, may not be reproduced in any form or by any means, electronic or mechanical, including photocopying, recording or any information storage and retrieval system now known or to be invented, without written permission from the publisher.

For photocopying of material in this volume, please pay a copying fee through the Copyright Clearance Center, Inc., 222 Rosewood Drive, Danvers, MA 01923, USA. In this case permission to photocopy is not required from the publisher.

For any available supplementary material, please visit
https://www.worldscientific.com/worldscibooks/10.1142/13689#t=suppl

Desk Editors: Nimal Koliyat/Amanda Yun

Typeset by Stallion Press
Email: enquiries@stallionpress.com

© 2024 World Scientific Publishing Company
https://doi.org/10.1142/9789811286711_fmatter

About the Editors

ArulMurugan Ambikapathi (S'02-M'11-SM'19) received a B.E. degree in electronics and communication engineering from Bharathidasan University, India, in 2003, an M.E. degree in communication systems from Anna University, India, in 2005, and the Ph.D. degree from the Institute of Communications Engineering (ICE) at National Tsing Hua University (NTHU), Taiwan, in 2011. He is currently a data science manager at Lam Research, Singapore, responsible for advanced deep learning computer vision and inverse design applications in semiconductor manufacturing. Prior to that, he was a group lead and scientist in Deep Learning 2.0/Machine Intellection, Institute of Infocomm Research, a research wing of the Agency for Science, Technology and Research, Singapore, from 2018 to 2022, focusing on industrial research and applications. Earlier, he was a team lead and senior algorithm engineer at Utechzone Co. Ltd., Taipei, Taiwan, from September 2014 to June 2018, where he developed one of the first AI-based defect inspection machines for manufacturing sectors. Prior to that, he was a postdoctoral research fellow with ICE at NTHU, from September 2011 to August 2014. His earlier research expertise includes convex optimization, biomedical and hyperspectral image analysis. His current research and application interests are in advanced computer vision, Bayesian optimization and online/continual learning (theory and applications).

Haytham M. Fayek received the B.Eng. (Hons) and M.Sc. degrees in 2012 and 2014, respectively, from the Petronas University of Technology, Malaysia, and the Ph.D. degree from the Royal Melbourne Institute of Technology (RMIT University), Melbourne, Victoria, Australia, in 2019. He was a postdoctoral research scientist at Meta, Washington, USA, from 2018 to 2020. He joined the faculty at RMIT University in 2020.

Dr. Fayek is a senior lecturer at the School of Computing Technologies at RMIT University. His research interests are in machine learning, deep learning and machine perception. He is a senior member of the Institute of Electrical and Electronics Engineers (IEEE), a member of the Association for Computing Machinery (ACM), and a member of Engineers Australia.

Li Xiaoli is currently the department head and a senior principal scientist of the Machine Intellection (MI) department at the Institute for Infocomm Research (I2R), A*STAR, Singapore. In addition to his role at A*STAR, he also serves as an adjunct full professor at the School of Computer Science and Engineering at Nanyang Technological University, Singapore. With a diverse range of research interests, Xiaoli focuses on cutting-edge areas such as AI, data mining, machine learning and bioinformatics. His contributions to these fields are evident through his extensive publication record, boasting over 300 peer-reviewed papers, and the accolades he has received, including nine best paper awards.

Xiaoli is an influential figure in the AI community, taking on leadership roles that significantly shape the field's development. As editor-in-chief of the *Annual Review of Artificial Intelligence* and an associate editor for prestigious journals, including *IEEE Transactions* on *Artificial Intelligence* and *Knowledge and Information Systems*, he plays a vital role in advancing AI research and disseminating valuable knowledge. His expertise and reputation have also earned him prominent roles in various conferences on AI and data analytics, where he serves as conference chair and area chair. Beyond academia, Xiaoli possesses extensive industry experience, where he has proven his ability to lead research teams effectively. Through his leadership, he has successfully spearheaded over 10 R&D projects in collaboration with major industry players across diverse sectors, such as aerospace, telecom, insurance and airline companies.

Xiaoli holds the title of IEEE Fellow and a Fellow of the Asia-Pacific Artificial Intelligence Association. He has also been recognized as one of the world's top 2% scientists in the AI domain by Stanford University. His contributions to both academic research and real-world applications make him a top expert in the field, and his dedication to advancing AI technology has far-reaching impacts on various sectors and industries.

Savitha Ramasamy (Senior Member, IEEE) received the Ph.D. degree from Nanyang Technological University, Singapore, in 2011. Currently, she is a research group leader at the Institute for Infocomm Research, A*STAR, Singapore. Her research interests are in developing robust AI, with a special focus on lifelong learning and time series data analysis, and she has led the translation of these robust models for predictive analytics in real-world applications. Her contributions to data analysis and AI have been recognized in the inaugural 100 SG Women in Technology list.

Suresh Sundaram is currently an associate professor in the Department of Aerospace Engineering at the Indian Institute of Science, Bengaluru, India. He received the B.E. degree in electrical and electronics engineering from Bharathiyar University, Coimbatore, India, in 1999, and the M.E. and Ph.D. degrees in aerospace engineering from the Indian Institute of Science, Bengaluru, India, in 2001 and

2005, respectively. He was a postdoctoral researcher with the School of Electrical and Electronic Engineering, Nanyang Technological University, Singapore, from 2005 to 2007.

From 2007 to 2008, he was with the National Institute for Research in Computer Science and Control-Sophia Antipolis, Nice, France, as a research fellow of the European Research Consortium for Informatics and Mathematics. He was with Korea University, Seoul, South Korea, for a short period as a visiting faculty in industrial engineering. From January 2009 to December 2009, he worked in the Department of Electrical Engineering at the Indian Institute of Technology, Delhi, India, as an assistant professor. From 2015 to 2018, he worked as an associate professor in the School of Computer Science and Engineering at Nanyang Technological University. His research interests include flight control, unmanned aerial vehicle design, machine learning, optimization and computer vision.

About the Contributors

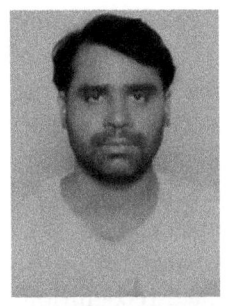

Ashish Mishra earned his bachelor's degree in mathematics from the University of Allahabad, Allahabad, India, in 2008. He completed his master's in computer science from the University of Hyderabad, India, in 2013. In 2021, he obtained his Ph.D. from the Department of Computer Science and Engineering at the Indian Institute of Technology Madras, India. His research focuses on applying zero-shot and continual learning in various domains, such as image classification, action classification and image retrieval.

Chandan Gautam has been working as a scientist at the Institute for Infocomm Research, Agency for Science, Technology and Research (A*STAR), Singapore, since 2022. Prior to joining A*STAR, he worked as a postdoctoral fellow at the Indian Institute of Science, Bengaluru, and ARTPARK, Bengaluru, India. He received the M.Tech. degree in information technology from the University of Hyderabad, Hyderabad, and the Institute for Development and Research in Banking Technology, Hyderabad, India, in 2014. Further, he received a Ph.D. degree in computer science and engineering from the Indian Institute of Technology Indore, India. His current research interests include open-world learning, continual learning and zero-shot learning.

Cuong V. Nguyen is an assistant professor at the Knight Foundation School of Computing and Information Sciences at Florida International University (FIU), USA. Before joining FIU, he worked as an applied scientist at Amazon Web Services and as a postdoctoral research associate at the University of Cambridge, UK. His research interests include probabilistic machine learning and artificial intelligence.

Davide Maltoni is a full professor in the Department of Computer Science and Engineering — DISI at the University of Bologna, Italy. He teaches machine learning and computer architectures at the University of Bologna. His research interests are in the areas of pattern recognition, machine learning, computer vision and computational neuroscience. Most of his applied research is in the field of biometric systems (fingerprint recognition, face recognition, hand recognition and performance evaluation of biometric systems). Davide Maltoni is co-founder and co-director of the Biometric Systems Laboratory (BioLab), which is internationally known for its research and publications in the field.

Gabriele Graffieti is currently a Ph.D. student at the Department of Computer Science and Engineering of the University of Bologna, Italy. His research is focused on continual learning and generative models, with a particular focus on continual learning from streaming data and generative models. He previously received a master's degree in computer science and engineering *cum laude* from the same university.

 **Hong Ren Wu** received the B.Eng. and M.Eng. degrees from the University of Science and Technology Beijing (formerly Beijing University of Iron and Steel Technology), China, in 1982 and 1985, respectively, and the Ph.D. degree in electrical and computer engineering from the University of Wollongong, Wollongong, New South Wales, Australia, in 1990.

From 1982 to 1985, he was an assistant lecturer with the Department of Industrial Automation at the University of Science and Technology Beijing. He joined the Department of Robotics and Digital Technology, Chisholm Institute of Technology, Caulfield, Victoria, Australia, as a lecturer, and then joined the Faculty of Information Technology, Monash University, Clayton, Victoria, Australia, in 1990, where he worked as a lecturer from 1990 to 1992, a senior lecturer from 1992 to 1996, and an associate professor of digital systems from 1997 to 2005. He was a professor of visual communications engineering with the Royal Melbourne Institute of Technology (RMIT University), Melbourne, Victoria, Australia, from February 2005 to November 2020, and concurrently worked as the head of Computer and Network Engineering from February 2005 to January 2010. He is currently an honorary professor at RMIT University. He has authored or coauthored articles that have appeared in refereed journals. He is a co-editor of the book *Digital Video Image Quality and Perceptual Coding* (Taylor and Francis, 2006). His research interests include signal processing, video and image processing and enhancement, perceptual coding of natural and medical images, digital video coding, compression and transmission, and digital picture quality assessment.

Dr. Wu was a guest editor for the Special Issue on Multimedia Communication Services of *Circuits, Systems and Signal Processing*, the Special Issue on Quality Issues on Mobile Multimedia Broadcasting of the *IEEE Transactions on Broadcasting*, and the Special Issue on QoE Management in Emerging Multimedia Services of the *IEEE Communications Magazine*.

Kanagasabai Rajaraman received his B.Eng. degree in 1989 from Anna University, India, and M.Eng. and Ph.D. degrees from the Indian Institute of Science, Bengaluru, India, in 1991 and 1997, respectively. Since 1997, he has held various R&D positions, with a focus on use-inspired research in data science, and created several award-winning data-driven AI technologies, many of which have been commercialized and led to successful spinoffs. He has published in many top peer-reviewed journals and conferences, served on the program committees of leading international conferences, and also chaired several international events on AI, data science and analytics. Dr. Kanagasabai is currently a senior scientist at the Institute for Infocomm Research (I2R), A*STAR, Singapore. His research interests include deep learning, NLP, explainable AI and semantic computing.

Lorenzo Pellegrini is a postdoc at the University of Bologna (UniBo), Italy. He received his Ph.D. in computer science and engineering at UniBo in 2022, with a dissertation on "Continual Learning for Computer Vision Applications." He has been working as a teaching tutor for the machine learning course offered as part of the computer science and engineering master's degree at UniBo since 2018. His main field of interest is continual lifelong learning, with a focus on the theoretical and practical aspects of handling fine-grained and long-tailed sequences of experiences. He is a lead maintainer for the Avalanche open-source continual learning library. In 2021, he worked as a research intern at Facebook AI Research. He served as an organizer of the 3rd and 4th challenges held at the CLVISION workshop at the IEEE Conference on Computer Vision and Pattern Recognition (CVPR) 2022 and 2023.

About the Contributors xiii

Pranshu Ranjan Singh received his M.Tech. degree in knowledge engineering from the National University of Singapore, Singapore, in 2019. Prior to that, he received his B.Tech. degree in computer science and engineering from the PDPM Indian Institute of Information Technology, Design and Manufacturing, Jabalpur, India, in 2017. He is currently a senior data scientist at Lam Research, Singapore. Previously, he worked as a research engineer at the Institute for Infocomm Research, a research wing of the Agency for Science, Technology and Research, Singapore, from 2019 to 2022. His research interests are in computer vision, deep learning, the design of experiments, image generation and continual learning.

Richard E. Turner is a professor of machine learning in the Department of Engineering at the University of Cambridge, UK. He is also a Bye fellow at Christ's College and a visiting researcher at Microsoft Research Cambridge. His research interests include probabilistic machine learning and machine learning for weather and climate.

Saisubramaniam Gopalakrishnan received his M.Tech. degree in knowledge engineering from the National University of Singapore, Singapore, in 2019. Prior to that, he received his B.E. degree in computer science and engineering from Sri Venkateswara College of Engineering (affiliated with Anna University), Tamil Nadu, India, in 2015. He is currently a senior research engineer at Quantiphi Analytics Solutions Pvt. Ltd., India. Previously, he worked as a research engineer at the Institute for Infocomm Research, a research wing of the Agency for Science, Technology and Research, Singapore, from 2019 to 2022. His research interests include incorporating factual evidence and alignment in large language models, continual learning, computer vision, the design of experiments and deep learning in general.

Sethupathy Parameswaran is currently working as a research engineer at the School of Computing and Information Systems at Singapore Management University, Singapore. He received his M.Tech. in aerospace engineering from the Indian Institute of Science (IISc), Bengaluru, India. He was a project assistant at the Artificial Intelligence and Robotic Lab (AIRL) at IISc from November 2020 to October 2022. His research interests include zero-shot learning, continual learning and out-of-distribution detection.

Siddharth Swaroop is a postdoctoral researcher in the Department of Computer Science at Harvard University, USA, in the Data to Actionable Knowledge group. Before that, he completed his Ph.D. at the University of Cambridge, UK. He works on probabilistic machine learning, approximate inference and applications to human–computer interaction.

Thang D. Bui is a lecturer in machine learning at the School of Computing, Australian National University (ANU), Australia. Prior to joining ANU, he was a lecturer at the University of Sydney and a research scientist at Uber AI. His research interests include probabilistic modeling and inference, deep learning and sequential decision-making.

Theivendiram Pranavan is a teaching assistant and a Ph.D. candidate in the Department of Computer Science at the National University of Singapore, Singapore. His research interests primarily focus on computer vision and machine learning. Notably, his research work has been published in renowned conferences, demonstrating the quality and significance of his contributions. Additionally, he has been recognized with a teaching award, highlighting his exceptional ability to effectively communicate complex concepts. Pranavan's academic prowess is evidenced by his achievement as one of the top 10 students in the Sri Lankan Advanced Level exams in the physical science stream. He earned his bachelor's degree from the University of Moratuwa, Sri Lanka, further solidifying his foundation in computer science and related disciplines.

Vincenzo Lomonaco is an assistant professor at the University of Pisa, Italy, where he teaches courses on artificial intelligence and continual learning. Currently, he also serves as co-founding president and lab director at ContinualAI, a non-profit research organization and the largest open community on continual learning for AI, a co-founding board member at AI for People, and a proud member of the European Lab for Learning and Intelligent Systems (ELLIS). In Pisa, he works within the Pervasive AI Lab and the Computational Intelligence and Machine Learning Group, which is also part of the Confederation of Laboratories for Artificial Intelligence Research in Europe (CLAIRE). Vincenzo is a task leader of two main European projects and a principal investigator of several industrial research contracts with companies such as Meta, Intel, Leonardo S.p.A. and SEA Vision S.r.l.

Yingzhen Li is a lecturer (equivalent to an assistant professor in the US) in machine learning at the Department of Computing, Imperial College London, UK. Before that, she was a senior researcher at Microsoft Research Cambridge. She received her Ph.D. in engineering from the University of Cambridge, UK. She works on probabilistic machine learning methods with applications to Bayesian deep learning and deep generative models.

© 2024 World Scientific Publishing Company
https://doi.org/10.1142/9789811286711_fmatter

Contents

About the Editors	v
About the Contributors	ix
List of Figures	xix
List of Tables	xxxi
1. **Introduction**	1
Chandan Gautam, Savitha Ramasamy, Suresh Sundaram and Li Xiaoli	
2. **Architectural Approaches to Continual Learning**	9
Haytham M. Fayek and Hong Ren Wu	
3. **Growing RBM on the Fly for Unsupervised Representation toward Classification and Regression**	25
Savitha Ramasamy, ArulMurugan Ambikapathi and Kanagasabai Rajaraman	

4. **Lifelong Learning for Deep Neural Networks with Bayesian Principles**   51

Cuong V. Nguyen, Siddharth Swaroop, Thang D. Bui, Yingzhen Li and Richard E. Turner

5. **Generative Replay-Based Continual Zero-Shot Learning**   73

Chandan Gautam, Sethupathy Parameswaran, Ashish Mishra and Suresh Sundaram

6. **Architect, Regularize and Replay: A Flexible Hybrid Approach for Continual Learning**   101

Vincenzo Lomonaco, Lorenzo Pellegrini, Gabriele Graffieti and Davide Maltoni

7. **Task-Agnostic Inference Using Base–Child Classifiers**   123

Pranshu Ranjan Singh, Saisubramaniam Gopalakrishnan, Savitha Ramasamy and ArulMurugan Ambikapathi

8. **Flashcards for Knowledge Capture and Replay**   163

Saisubramaniam Gopalakrishnan, Pranshu Ranjan Singh, Haytham M. Fayek, Savitha Ramasamy and ArulMurugan Ambikapathi

9. **Reliable AI-Based Decision Support System for Chest X-Ray Classification Using Continual Learning**   197

Theivendiram Pranavan and ArulMurugan Ambikapathi

Bibliography   217

Index   237

© 2024 World Scientific Publishing Company
https://doi.org/10.1142/9789811286711_fmatter

List of Figures

1.1 Lifelong vs. continual vs. online learning. 2
2.1 Progressive neural networks for three tasks. 13
2.2 Overview of progressive learning for three tasks. A curriculum strategy is used to select a task (blue) from the pool of candidate tasks. Next, a model is trained to perform the selected task (blue), and the learned parameters remain constant thereafter. Then, the curriculum strategy is employed to select the subsequent task (purple). Afterward, new model parameters, denoted progressive blocks, which draw connections from the preceding layer in the block as well as the preceding layer in prior progressive block(s), are added and trained to perform the selected task (purple). Subsequently, after training the newly added progressive block to convergence, a pruning procedure is used to remove weights without compromising performance. Finally, the curriculum, progression and pruning procedures are repeated for the third task (green) and for all remaining task(s) thereafter (Fayek et al., 2020). 19

List of Figures

2.3 Accuracy of progressive learning vs. independent learning for all 11 image recognition tasks. The tasks are ordered according to the outcome of the curriculum procedure. C-10 denotes the CIFAR-10 task, and C-100-s indicates the sth CIFAR-100 task (Fayek *et al.*, 2020). 22

3.1 Network architecture and training phases: Online representation learning is performed in the first phase, wherein the network begins with zero neurons in the hidden layer and adds and/or adapts as the network learning progresses to derive a feature representation of the data in an unsupervised manner. The next phase performs supervised discriminative learning to associate the feature representation with the class labels. 30

3.2 The overall block diagram for the proposed online restricted Boltzmann machine. 31

3.3 MNIST: Reconstruction error and addition of neurons to the generative RBM network through online learning: (a) reconstruction error of samples 1–30,000; (b) growth of the network with samples 1–30,000. 37

3.4 Choice of threshold, E_n, of O-RBM for the MNIST data set. 38

3.5 MNIST classification: Average number of hidden layer neurons associated with each class of the MNIST data set, with the standard deviation across 50 trials. 39

3.6 MNIST classification: Reconstruction of the handwritten digits using the online generative network. The odd-numbered rows represent the original image, while the even-numbered rows are the images reconstructed by the RBM network trained with the O-RBM. 40

3.7 The evolution of add threshold and the network hidden size, with streaming data in the UK region. 46

3.8 Incremental learning of O-RBM using data from multiple regions: The green line represents the end of training using data from stations in the Gulf of Mexico, and the blue line represents the end of training using data from stations in the Korean region. 47

3.9 Transferring representations from source regions to the Irish region: RMSEs of DBN vs. O-RBM. 48

List of Figures

4.1 Schematic diagrams of multi-head networks, including both the probabilistic graphical model (left) and network architecture (right), reproduced from Nguyen et al. (2018): (a) A multi-head discriminative model showing how network parameters might be shared. The lower-level network is parameterized by the variables $\boldsymbol{\theta}^S$ and is shared across multiple tasks. Each task $t \in \{1, 2, \ldots, T\}$ has its own "head network" $\boldsymbol{\theta}_t^H$ mapping onto the outputs from a common hidden representation. The full set of parameters is therefore $\boldsymbol{\theta} = \{\boldsymbol{\theta}_{1:T}^H, \boldsymbol{\theta}^S\}$. (b) A multi-head generative model (see Section 4.3.3 for details) with shared network parameters. For each task t, the head networks $\boldsymbol{\theta}_t^H$ generate intermediate-level representations from the latent variables $\mathbf{z}_t$. 55

4.2 The active units learned for Split MNIST without coresets. Rows correspond to stages of lifelong learning: (left) Means, (center) Variances, (right) Output weights for each task's two classes. Exactly the same effect is observed when incorporating coresets. 67

4.3 Numbers of active units per hidden layer after each task in Permuted MNIST without coresets. Exactly the same effect is observed when incorporating coresets. 68

5.1 Network architecture of CVAE used in the proposed GRCZSL method: The input feature x and the respective attribute vector a_y are concatenated and passed through a fully connected layer (L_1) of 512 units, followed by dropout with a dropout rate of 0.3 and passed through the fully connected layer (L_2) of 512 units. From (L_2), μ_z and Σ_z are further obtained individually via another fully connected layer of dimension 50. z is sampled from the variational distribution $\mathcal{N}(\mu_x, \Sigma_x)$. The sampled z concatenated with a_y is passed to a fully connected layer (L_3) of 1,024 units and the original x is then reconstructed. In this architecture, we use ReLU as an activation function at all layers, except the outputs of encoder and decoder, where we use a linear activation. 83

5.2	Flow diagram of GRCZSL: $E^{(t)}$ and $D^{(t)}$ are the encoder and decoder for the tth task, which learns from the replay data and also performs CZSL. $D^{(t-1)}$ is the decoder of the $(t-1)$th task, which generates sample for generative replay.	84
5.3	CZSL setting by Wei *et al.* (2020).	89
5.4	Fixed and dynamic CZSL settings.	90
5.5	t-SNE plot of 10 random classes of the CUB dataset for ResNet-101 features and synthetic features generated through CVAE.	95
5.6	Harmonic mean for the CUB dataset over 20 tasks for fixed and dynamic CZSL settings.	98
5.7	Per class unseen accuracy for the CUB dataset over 20 tasks for fixed and dynamic CZSL settings.	98
5.8	Per class unseen accuracy as per task importance for the CUB dataset over 20 tasks for fixed and dynamic CZSL settings.	99
5.9	The impact of the task importance on the CUB dataset in terms of mH, mSA and mUA for both settings.	99
6.1	A Venn diagram of some of the most popular CL strategies: CWR (Lomonaco and Maltoni, 2017), PNN (Rusu *et al.*, 2016a), EWC (Kirkpatrick *et al.*, 2016), SI (Zenke *et al.*, 2017), LWF (Li and Hoiem, 2016), ICARL (Rebuffi *et al.*, 2017b), GEM (Lopez-Paz and Ranzato, 2017), FearNet (Kemker and Kanan, 2018), GDM (Parisi *et al.*, 2018), ExStream (Hayes *et al.*, 2018a), Pure Rehearsal, GR (Shin *et al.*, 2017), MeRGAN (Wu *et al.*, 2018) and AR1 (Maltoni and Lomonaco, 2019). Rehearsal and generative replay upper categories can be seen as subsets of replay strategies.	104
6.2	Architectural diagram of ARR (Pellegrini *et al.*, 2019).	110
6.3	Accuracy on iCIFAR-100 with 10 batches (10 classes per batch). Results are averaged over 10 runs: For all the strategies, hyperparameters have been tuned on run 1 and kept fixed in the other runs. The experiment on the right, consistent with the CORe50 test protocol, considers a fixed test set, including all 100 classes, while	

on the left we include in the test set only the classes encountered so far (analogously to the results reported by Rebuffi et al., 2017b). Colored areas represent the standard deviation of each curve. Better viewed in color (Maltoni and Lomonaco, 2019). 112

6.4 Accuracy results on the CORe50 NICv2-391 benchmark of ARR(α = $pool6$), ARR(λ = 0.0003), DSLDA, iCaRL, ARR(RM_{size} = 1500, α = $conv5\_4$) and ARR(RM_{size} = 1500, α = $pool6$). Results are averaged across 10 runs in which the batch order is randomly shuffled. Colored areas indicate the standard deviation of each curve. As an exception, iCaRL was trained only on a single run given its extensive run time (∼14 days). 114

6.5 ARR with latent replay (RM_{size} = 1500) for different choices of the latent replay layer. Setting the replay layer at the "images" layer corresponds to native rehearsal. The saturation effect which characterizes the last training batches is due to the data distribution in NICv2-391 (see Lomonaco et al., 2019b): in particular, the lack of new instances for some classes (that already introduced all their data) slows-down the accuracy trend and intensifies the effect of activations aging. 117

6.6 Comparison of main ARR configurations on CORe50 NICv2-391 with different external memory sizes (RM_{size} = 500, 1000, 1500 and 3000 examples). 119

6.7 ARR implementation in Avalanche. Given a set of hyperparameters, ARR can be instantiated and properly configured to be tested on a large set of benchmarks already available in Avalanche. 120

7.1 Block diagram for BCCs. This depicts the various components of BCC and describes the method for performing task-incremental classification. 129

7.2 t-SNE visualization for reference points R_{T_k} and its effect on the trained base classifier's LS of the MNIST dataset. 131

7.3 Block diagram for the base classifier: A weighted combination of cross-entropy and clustering loss is used for optimizing the base classifier network. 132

xxiv   *List of Figures*

7.4   Selected points (red) shown along with training samples (blue) and the class mean (green) using t-SNE visualization for class 0 (airplane) of the Cifar10 dataset. 135

7.5   Points lying on the boundaries (shown in different colors) of each class cluster are selected boundary points for the Cifar10 dataset. Labels 0–9 indicate the 10 classes in Cifar10, and labels 10–19 correspond to the selected boundary points of those classes. 135

7.6   Univariate combinations sampling method. Blue and orange blobs indicate the training data points and selected samples, respectively. 137

7.7   Outermost cluster sampling method. Blue and orange blobs indicate the training data points and selected samples, respectively. 138

7.8 t-SNE projections-based convex/concave hull sampling method. Selected points (red) using convex hull (*left*) and concave hull (*right*) from the t-SNE classification LS of the Cifar10 dataset class 0 (*airplane*). 140

7.9   Block diagram for the LS reconstructor. The LS reconstructor model is optimized using MAE loss, and the training set is formed by combining boundary points sample from all the tasks observed until now. 141

7.10 Block diagram for the child classifier. This model appends the output layer of a specific base classifier (classification head) on top of the frozen LS reconstructor model to perform inference. 142

7.11 ACC scores computed after observing each task of Split-Cifar10 for the "without Task ID" scenario, across all discussed methods. 150

7.12 BWT scores computed after observing each task of Split-Cifar10 for the "without Task ID" scenario, across all discussed methods. 151

7.13 ACC scores computed after observing each task of Split-Cifar10 for the "with task ID" scenario, across all discussed methods. 151

7.14 BWT scores computed after observing each task of Split-Cifar10 for the "with task ID" scenario, across all discussed methods. 152

7.15	t-SNE visualization for Cifar10 (left) and MNIST (center) classification obtained using respective base classifiers on corresponding test sets. t-SNE visualization for combined Cifar10 (indices 0–9) and MNIST (indices 10–19) classification obtained from the child classifier (right).	152
7.16	Test accuracy scores for Cifar10 and MNIST after observing both tasks, across different sampling sizes using random sampling.	157
7.17	Test accuracy scores for Cifar10 and MNIST after observing both tasks, across different sampling sizes using farthest distance sampling.	158
8.1	Flashcards for knowledge capture and replay: step (a) train the autoencoder (AE) for Task T_1; step (b) construct flashcards from frozen AE using maze patterns via recursive passes; step (c) replay using flashcards on a new network to remember Task T_1; step (d) replay using flashcards F_t (containing consolidated knowledge from task T_1 to T_t) while training for Task T_{t+1}.	165
8.2	A sample of three flashcard (F_t) constructions over recursive iterations, with Cifar10 as an example: Difference in MAE between successive iterations with iteration number and t-SNE of latent space in 2D. It can be observed that the initial (raw) input maze patterns adapt to the texture captured by the model as the number of recursive iterations increases (iteration $r = 10, 20$). Also, the corresponding latent space clusters get closer. However, on further repeated passes ($r \geq 50$), the reconstruction deteriorates/smooths out (as MAE keeps reducing), and the two clusters drift apart again (Property 8.1 is satisfied but Property 8.2 is not). Best viewed in color.	168
8.3	Transformation from maze patterns (D_m) into flashcards (F_t) for tasks: MNIST, Fashion MNIST and Cifar10. . .	169
8.4	Analysis on iterations r for flashcard construction, demonstrated using SVHN and Cifar10. Row 1 portrays metrics corresponding to the change on passing multiple random maze inputs in iterative fashion through	

an autoencoder. The orange curve depicts the trend of FLSD between intermediate input at iteration t and original data samples, and the dark-blue curve shows a constant decrease in reconstruction error between $t-1$ and t across different iterations for each dataset. The light-blue line serves as a reference MAE on the original data. Property 8.1 is satisfied when the blue curve goes below ϵ_1 (marked by the dotted red line), and Property 8.2 is satisfied at the first minima of FLSD, which occurs around $10 \leq r \leq 20$, irrespective of dataset. Row 2 compares the autoencoder error (MAE) between test data and its reconstruction when trained on flashcards constructed from different iterations. This again confirms the acceptable range of r and that flashcard construction is *not critically sensitive* to the choice of r within the range. 171

8.5 Increase in number of flashcards (denoted by fc-) closely follows the improvement observed by training original (coreset) samples, as observed in (i) Cifar10 and (ii) MNIST. 173

8.6 Visual comparison of different initializations used for the construction of flashcards: Maze, Gaussian noise and next task. Flashcards constructed after training on a particular dataset subject to initialization are presented. Only maze-based flashcards have diversity in terms of shape and texture and perform best due to their ability to capture activations at edges. Gaussian noise results in repetitions. Flashcards from new task initialization are dependent on the random subset images' coverage of old tasks' activations and lead to inconsistent results (Table 8.2). 174

8.7 Reconstruction error of two networks, one trained on original ImageNet 256 × 256 resolution 1M images (gray), and the second trained only on flashcards 256 × 256 resolution (orange). The x-axis denotes the percentage of samples added, and as more flashcards are introduced, the error in the second network approaches closer to that of the first. 175

8.8 Comparison between different methods for continual reconstruction using Sequence5. The figure shows the impact of forgetting after the end of the *last* task, i.e., Omniglot (all intermediate results are in Supplementary). Each row presents a dataset/task. SFT (left) is not able to reconstruct, even when presented with Omniglot, because of (i) catastrophic forgetting and (ii) unsuitable transfer learned weight initialization. Continual VAE (center) fares better but has artifacts and color loss in the reconstructions. Flashcards (right) remember all previous tasks. 187

8.9 Continual reconstruction on naive/sequential fine-tuning (SFT). 188

8.10 Continual reconstruction using episodic memory: Coreset 500. 188

8.11 Continual reconstruction using VAE trained exclusively for CL. 189

8.12 Continual reconstruction using AE + VAE as generative replay. 189

8.13 Continual reconstruction using flashcards. 190

8.14 Continual learning for reconstruction. Individual graphs for different methods show the variation of test MAE on the current task dataset after observing the data for a sequence of tasks. The table shows the test MAE for the joint training (JT) method. The reported values in the graph and table were obtained over five experimental runs. The standard deviation is quite small and is not displayed on the graphs to avoid clutter. 190

8.15 Continual reconstruction on the UC Merced land use 256 px dataset. The line plot shows reconstruction error averaged from the first to the current task, as new tasks are introduced. SFT and coreset oscillate occasionally as new tasks are encountered, e.g., beach (blue texture) after agriculture and baseball ground (green), forest after series of buildings (gray), golf course after freeway, overpass after park, etc. Despite minimal capacity among other methods, flashcards have a relatively stable transition (dotted horizontal line shows peak error) and low error for all tasks. 191

8.16	Continual denoising scenario. Shown in the figure is the effect of noise applied and the reconstruction of Sequence5 using flashcards.	193
9.1	Proposed AI framework: In this framework, we feed the given two datasets, and we continually learn a binary classification of whether a disease is there or not. In certain cases of inference where the machine learning model is not confident about the classification decision, the model seeks medical practitioner's assistance for expert labeling.	200
9.2	CheXpert dataset: Sample images. Here, the two images show the frontal and lateral views of the same patient.	201
9.3	Example input images in a binary classification problem of a disease: (a) and (b) images with the disease pneumothorax, and (c) and (d) are those without the disease.	203
9.4	Mimic-CXR images.	204
9.5	ResNet-50-based architecture. The weights are pretrained from the Imagenet dataset. The final layer is modified to solve the binary classification problem. The same architecture is used for binary classification problems.	204
9.6	The plots show individual and cross performance. The left plot (a) shows a model trained using the CheXpert dataset and evaluated on both datasets. Similarly, the plot on the right (b) is trained on MIMIC-CXR.	206
9.7	EWC (Aich, 2021), which finds a parameter set in the overlapping region. It builds on the assumption that the solution spaces of various tasks overlap.	207
9.8	OOD detection framework: Training. Given the set of training images, two isolation forest models are trained for OOD detection (as an anomaly case) over latent space representations obtained via a pretrained DL model.	210

9.9	OOD detection framework: Inference. The activations obtained from a pretrained DL model for a given test image is used to obtain anomaly scores from two trained (healthy and disease) isolation forest models, thereby determining whether the test image is OOD or not.	212
9.10	Lateral view images are fed to the OOD detector. The proposed OOD detector can detect all 10 samples which are misclassified as OOD. Here, if two OOD scores are negative, the sample does not belong to either of the distributions.	213
9.11	OOD detection with Gaussian blur and shot noise.	213
9.12	OOD with augmentation.	214
9.13	OOD with different classes of images.	214

# List of Tables

3.1	Description of the credit scoring data sets.	42
3.2	Performance analysis of benchmark data sets: Credit scoring.	43
3.3	Data characteristics.	45
3.4	Performance study results of wave height prediction: Test data in the individual source regions.	46
4.1	Final average test accuracy on Split MNIST for various methods. Methods with an asterisk (*) use some sort of episodic memory. Results marked with a double asterisk (**) were taken or read from a graph in the respective papers.	64
4.2	Final average test accuracy on Permuted MNIST for various methods (results taken from respective papers). A hidden layer size of $\{n_1, n_2\}$ indicates two hidden layers, the lower hidden layer having n_1 hidden units and the upper hidden layer having n_2 units (followed by a softmax over the 10 MNIST classes). Methods with an asterisk (*) use some sort of episodic memory. Results marked with a double asterisk (**) were read from a graph in the source paper. Results with a dagger (†) are from Chaudhry et al. (2018b).	65

5.1	Comparison between our work and the study by Wei et al. (2020).	80
5.2	Standard split of ZSL datasets.	87
5.3	Hyperparameters for GRCZSL in fixed CZSL setting.	88
5.4	Hyperparameters for GRCZSL in dynamic CZSL setting.	88
5.5	mSA, mUA and mH values are provided for fixed CZSL setting. The bold face represents the best results in the table.	94
5.6	mSA, mUA and mH values are calculated for dynamic CZSL setting. The bold face represents the best results in the table.	96
5.7	The table provides results of the CZSL methods at the last task when the train–test split is identical to the offline case, where all training and testing data are available at once. We also provide the results of the offline method, which is basically an upper bound for the CZSL methods available in the table.	97
6.1	Final accuracy on ImageNet-1000 following the benchmark of Masana et al. (2020) with 25 experiences composed of 40 classes each. For each method, a replay memory of 20,000 examples is used (20 per class at the end of training). Results for other methods reported by Masana et al. (2020).	116
6.2	Computation–storage–accuracy trade-off with latent replay at different layers of a MobileNetV1 ConvNet trained continually on NICv2-391 with $RM_{\text{size}} = 1500$.	118
7.1	Results for the Split-Cifar10 experiment. BCC outperforms other methods for the "without task ID" scenario.	149
7.2	Results for the Cifar10–MNIST experiment. BCC outperforms other methods for the "without task ID" scenario.	153

7.3	Results for various sampling methods used with BCC for the Cifar10–MNIST experiment. The sampling fraction p is set at 0.1, or 10%, for all methods.	155
7.4	Results for various sampling methods used with BCC for the Cifar10–MNIST experiment. The sampling fraction p is set to 0.05 or 5% for all methods.	156
7.5	Results for different sample sizes when using the disjoint univariate combinations sampling method with BCC for the Cifar10–MNIST experiment.	158
7.6	Results for different sample sizes when using the t-SNE projections-based convex/concave hull sampling method with BCC for the Cifar10–MNIST experiment.	159
7.7	Results for different sample sizes when using the farthest from each point sampling method with BCC for the Cifar10–MNIST experiment.	160
8.1	Benchmarking effect of iterations for reconstruction of different datasets, using 50,000 flashcards.	172
8.2	Maze patterns provide the best reconstruction MAE among the different initializations used for construction of flashcards. Experiments run on Sequence5 continual reconstruction.	173
8.3	Reconstructions from two separate autoencoders, one trained on original data (Network1) and the other trained on flashcards (Network2). The metrics indicate that flashcards are sufficient alternatives for learning the original data. Weighted alpha (Martin and Mahoney, 2020) is based on HT-SR theory. Alpha values are closer for the trained networks, which indicate a similarity between the two network weights. The reported values are averaged over five runs.	175
8.4	Building autoencoder AE2 (smaller and larger arch.) using flashcards obtained from AE1 trained on the Cifar10 dataset. The column "Test original" shows test MAE when these architectures are trained using the original Cifar10 dataset. The column "Test AE2" refers to test MAE on Cifar10 when flashcards obtained from AE1	

	are used to train AE2 for reconstruction. Better performance on both the AE2-Smaller and AE2-Larger architectures showcases transfer using flashcards.	176
8.5	Building a classifier using the reconstructions from flashcard-trained network. The reported accuracies are averaged over five runs. Cifar10 accuracy is lower due to a relatively higher autoencoder reconstruction error.	177
8.6	Comparison between different continual learning replay strategies.	179
8.7	Architecture selection for autoencoder (AE). Several AE architectures were trained on the Cifar10 dataset in order to compare the performance of flashcards for reconstruction. Various details about the architecture, such as Model Params. (number of trainable weights and biases), Latent Space (size of latent space/bottleneck layer and its reduction rate versus image space), Num. Blocks (number of convolution + pooling blocks in encoder), and Num. Filters (number of filters in convolution layers) are also provided. Original MAE is the Cifar10 test MAE on AE trained using the Cifar10 train dataset. Flashcards MAE is the Cifar10 test MAE on the AE trained using the flashcards obtained from the given AE. The reported standard deviation for the scores was obtained over five experimental runs.	182
8.8	Continual reconstruction for homogeneous tasks (each task T_k specifies up to k classes from the same dataset that have been observed) shows minimal/no forgetting, with examples showcasing reconstruction error for Cifar10 and Caltech101. On the contrary, there is a strong indication that knowledge from the past improves learning newer tasks (forward transfer).	184
8.9	Sequence3 order: MNIST, Fashion MNIST and Cifar10 as three tasks. Tasks are added incrementally, and MAE is computed on each dataset after the current task is completed.	185
8.10	Sequence3 order: Fashion MNIST, Cifar10 and MNIST as three tasks.	185
8.11	Sequence3 order: Cifar10, MNIST and Fashion MNIST as three tasks.	186

8.12	Continual learning for Sequence5 reconstruction and denoising, compared with different methods in terms of error observed in MAE and BWT on the test set. Additionally compared are the network capacity (N/w in MB) and external memory (Mem in MB) required by each method.	192
8.13	Comparison of different methods for task-incremental learning using Sequence3 tasks (Cifar10–MNIST–Fashion MNIST). The reported values are accuracies computed at the end of task T.	194
8.14	ST-NIL classification on Cifar10 using the settings described by Tao *et al.* (2020).	195
8.15	ST-NIL classification on Cifar10 on the modified brightness and saturation setting with mean and standard deviation over five runs.	196
9.1	CheXpert and MIMIC-CXR datasets summary.	201
9.2	Deep neural network settings.	205
9.3	CheXpert dataset performance.	205
9.4	MIMIC-CXR dataset performance.	205
9.5	MIMIC model tested on CheXpert.	206
9.6	CheXpert model tested on MIMIC.	206
9.7	CL strategy and results.	208

© 2024 World Scientific Publishing Company
https://doi.org/10.1142/9789811286711_0001

Chapter 1

Introduction

Chandan Gautam[*,‡], Savitha Ramasamy[*,§], Suresh Sundaram[†,∥]
and Li Xiaoli[*,¶]

[*]*Institute for Infocomm Research, Agency for Science,
Technology and Research (A*STAR), Singapore*
[†]*Indian Institute of Science, Bangalore, India*
[‡]*gautam_chandan@i2r.a-star.edu.sg*
[§]*ramasamysa@i2r.a-star.edu.sg*
[∥]*vssuresh@iisc.ac.in*
[¶]*xlli@i2r.a-star.edu.sg*

1.1 Introduction

Over the past two decades, machine learning methods have exhibited tremendous performance and outperformed humans in a variety of tasks, such as voice recognition, object recognition and various games. Nevertheless, machine intelligence is far behind human intelligence in solving various tasks. Throughout their lives, humans and animals have the potential to acquire, fine-tune, accumulate and utilize the acquired knowledge and abilities from one problem to solve another. This whole process of learning is broadly known as lifelong or continual learning, in which a machine leverages the knowledge learned in the past to learn a new task. Overall, lifelong or continual learning aims to mimic humans' continuous learning process by incrementally learning a sequence of tasks and accumulating knowledge,

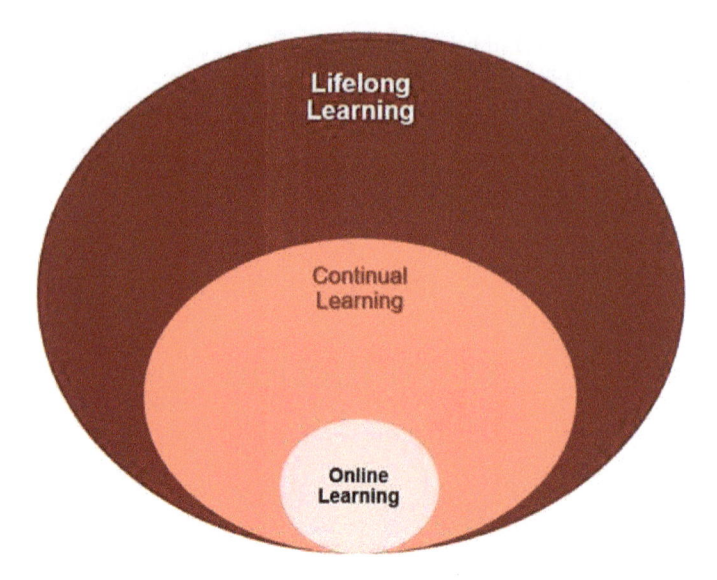

Fig. 1.1: Lifelong vs. continual vs. online learning.

which is adapted for future learning. However, most of the studies on lifelong or continual learning focused on the offline training and testing approach, i.e., there is no learning after deploying it. It means that the model doesn't learn while performing any tasks. When the model learns in an online manner, i.e., after deployment, this way of learning is known as on-the-job learning or online continual/lifelong learning. Without this kind of learning ability, it's difficult to envisage a really intelligent machine. We will further discuss lifelong learning, continual learning and online learning, which are depicted in Fig. 1.1. Although lifelong and continual learning essentially signify the same thing, previous studies conducted under the two labels have focused on distinct elements of the same problem.

1.2 Lifelong Learning

Chen and Liu (2016) define lifelong learning as follows: Let a model learn from N sequential tasks, $T_1, T_2, \ldots, T_{N-1}, T_N$. After training the model with N tasks, when it encounters a T_{N+1}^{th} task, the model needs to leverage the accumulated knowledge from the knowledge base (KB) to learn the new T_{N+1}^{th} task. Here, KB consists of the accumulated knowledge from the past N tasks. After learning the

$(N+1)^{\text{th}}$ task, the model updates the KB with the knowledge gained from the $(N + 1)^{\text{th}}$ task. Overall, the goal of lifelong learning is to leverage the past accumulated knowledge to learn new knowledge.

1.3 Continual Learning

In the deep learning community, the term continual learning (CL) is more widely used than lifelong learning. CL's main goal is to eliminate catastrophic forgetting when a machine leaning model learns from never-ending non-stationary (locally independent and identically distributed, or i.i.d.) streaming data. This way of learning differs significantly from the typical machine learning method, where all data need to be available before training (i.e., stationary data) instead of streaming non-stationary data. CL mainly needs to tackle the following two challenges:

(1) *Catastrophic forgetting*: While learning from a continuous stream of data, a neural network (or any machine learning method) might forget the past accumulated knowledge as the weights of the network adjust according to the new distribution of tasks and perform inferiorly on the previously learned tasks. Overall, positive and negative backward transfers play a crucial role in catastrophic forgetting. Positive backward transfer reduces the impact of forgetting, and vice versa is true for negative backward transfer.

(2) *Intransigence*: When a model is unable to learn new knowledge, it is known as intransigence. It can be due to either model incapability or the model being highly regularized to maintain the knowledge gained from the prior tasks. Hence, if a model is highly regularized, then catastrophic forgetting will be less but intransigence will be more. On the contrary, if a model is weakly regularized, then intransigence will be less, but it will suffer from catastrophic forgetting. Hence, we need to develop a balanced model which can handle both issues efficiently. Overall, a model needs to do positive forward transfer to reduce intransigence.

1.3.1 *Evaluation settings in continual learning*

In CL, data are divided into multiple tasks and then streamed task-wise, one after the other, for training and testing the model.

Based on these tasks, the performance of CL is evaluated in the literature using the following settings:

(1) *Multi-head setting*: Task identity (or task description) is known during training and testing.
(2) *Single-head setting*: The multi-head setting needs a task identifier at the inference time, which is not practical in real time. To address this issue, two types of evaluation strategies are developed in a single-head setting, as follows:
   (a) *task-agnostic prediction*: when task identity is known during training but not during testing;
   (b) *task-free learning*: when task identity is neither known during training nor during testing

In CL settings, when we discuss multiple tasks, we make the implicit assumption that these tasks must be related; otherwise, the transfer of knowledge is not possible. For example, if one task involves classifying among digits and another involves vehicle classification, then neither forward nor backward transfer is possible in CL. In this case, CL needs to handle only one problem, i.e., catastrophic forgetting.

1.3.2 A brief overview of continual learning methods

Overall, CL methods can be broadly categorized into three categories:

(1) *Regularization-based methods*: These methods address catastrophic forgetting by imposing some constraints on the weight update on the network. Overall, methods developed based on regularization can be divided into two types: data-focused and prior-focused. In the first, knowledge distillation is deployed to transfer the knowledge from the model learned over the previous tasks to the current model, which is trained over the new task. In the second, parameters are penalized based on their importance, and prior knowledge is consolidated while learning the new task.
(2) *Replay-based*: Methods developed in this category either rely on the rehearsal of a subset of a raw sample or a synthetic sample. In the case of a raw sample, the model needs a tiny episodic memory

to store the subset from the incoming data stream by using sample selection techniques, such as reservoir, ring buffer, mean of features-based and k-means-based sampling. These stored samples are replayed from memory while the model is trained on the new tasks. Overall, the model is jointly trained over the current data as well as a subset of the previous data from the episodic memory, and this process alleviates the catastrophic forgetting of the model. Further, this episodic memory is deployed to store the gradient of the trained model over the previous tasks instead of storing the raw samples. Here, a constrained optimization is deployed under the constraint that the update of the model for the new task should not interfere with the previous task. This is an alternative solution to alleviating catastrophic forgetting.

Another way of handling catastrophic forgetting is based on the replay of synthetic samples, which is known as pseudo-rehearsal. Here, instead of storing the raw samples in the memory, a generative model, which is trained over the previous tasks, is used to generate synthetic samples for the previous tasks, and these synthetic samples are combined with the new task to train the model. It is to be noted that these rehearsal and pseudo-rehearsal methods were initially developed for shallow neural networks to mitigate catastrophic forgetting. Later, it will be developed for deep neural networks.

(3) *Parameter-isolation-based*: The methods belonging to this family handle catastrophic forgetting by assigning different parameters of the model to each task. These methods are broadly divided into two types based on the model architecture: fixed and dynamic. In the case of a fixed architecture, different parts of it are allocated for each task, and the previous part is masked out while training for a new task. In the case of dynamic architecture, we can grow the architecture by adding new parameters for the new task while the original parameters remain intact. Here, the majority of past work requires a task identifier at the inference time (multi-head setting); however, only a few studies have recently addressed this issue and developed it for a single-head setting using an expert gate.

Apart from the theoretical development in CL, where most of the work is for the classification task on the image data, it is also explored for different kinds of other tasks, like neural machine translation, cross-lingual modeling, text classification, natural language generation, question-answering, sentiment analysis, object detection and action recognition.

1.4 Online Learning in Open- and Closed-World Environments

In the literature of CL, most of the works still locally train/test on multiple tasks in an offline/batch manner. Once these CL methods are deployed after training, they do not perform any kind of learning while working on the job or task. However, on the contrary, humans learn most of the information on the job after initial training. Another learning method reported in the literature, known as online learning, can address this issue. In online learning, a model learns samples sequentially and updates itself as it receives a new sample. However, online learning-based methods are only suitable for the closed-world environment. However, on-the-job learning performs online learning as well as learning in an open environment. Most of the CL methods only work in a closed-world environment, where it is assumed that classes during testing are already seen during training. It means that a model cannot encounter anything new during testing. However, this is not true in real-world applications. For example, a self-driving car can encounter new objects (unseen classes) while driving, so the model needs to learn unseen classes on the job. This learning process is called learning in an open-world environment. Overall, on-the-job learning predominantly performs the following three steps:

(1) Detect the samples of unseen classes from the streaming data.
(2) Identify the unseen classes and gather the training data to add to the KB.
(3) After gathering a sufficient amount of training data for those unseen classes, send those samples to the model to learn incrementally, which is called class-incremental learning.

Here, labeling the unseen classes is another challenge which needs to be addressed. It can be mainly solved in three ways through self-supervised interactions with humans and the environment:

(1) *Trial and error*: The system takes action in the environment and observes the impact of the action. Based on this, the model gathers data for further training.
(2) *Asking the user*: The system communicates with humans by posing questions to them. Their responses can be used as supervisory data for further training.
(3) *Reasoning and imitation*: Through reasoning and imitation, the system infers class labels based on its prior knowledge, e.g., zero-shot learning.

In on-the-job learning, apart from unseen classes, a model also needs to identify new knowledge for the seen class and add it immediately to the KB for training. This aspect has been well addressed in the literature by using meta-learning, where the model performs online learning based on the following three queries: what to learn, when to learn and how to learn. In recent years, a few researchers have explored online CL for open-world environments and zero-shot learning. There is extensive scope in this setup which needs to be further explored.

© 2024 World Scientific Publishing Company
https://doi.org/10.1142/9789811286711_0002

Chapter 2

Architectural Approaches to Continual Learning

Haytham M. Fayek[*,†] and Hong Ren Wu[*,‡]

*School of Computing Technologies,
Royal Melbourne Institute of Technology (RMIT) University,
Melbourne, Australia
†haytham.fayek@ieee.org
‡henry.wu@rmit.edu.au

Abstract

Continual learning is the ability of a learning system to solve new tasks by utilizing previously acquired knowledge from learning and performing prior tasks without having significant adverse effects on the acquired prior knowledge. Numerous approaches were developed to achieve this ability while avoiding the well-known problem of catastrophic forgetting in neural networks. This chapter reviews a number of architectural approaches to continual learning in neural networks which tackle the problem by modifying the architecture of the neural network, for example, by adding new adaptive parameters to the model for each new task. The architectural paradigm for continual learning in neural networks can potentially completely eliminate the problem of catastrophic forgetting and maintain competitive performance, often at the expense of increased computational complexity.

2.1 Introduction

The standard paradigm in machine learning aims to learn a single task from data. Continual learning aims to learn tasks in a sequential manner (Thrun, 1996; Mitchell *et al.*, 2018). This leads to the problem of catastrophic forgetting in neural networks, where the performance of a model on prior tasks significantly deteriorates when presented with a new task to learn. Numerous paradigms were developed to overcome catastrophic forgetting in neural networks, as follows (Fayek, 2019).

The architectural paradigm sidesteps catastrophic forgetting by adding new adaptive parameters to the model for each new task, such as block-modular neural networks (Terekhov *et al.*, 2015), progressive neural networks (Rusu *et al.*, 2016b), residual adapters (Rebuffi *et al.*, 2017a), dynamically expandable networks (Yoon *et al.*, 2018), reinforced continual learning (Xu and Zhu, 2018a), continual learning via neural pruning (Golkar *et al.*, 2019), progressive learning (Fayek *et al.*, 2020) and efficient feature transformations (Verma *et al.*, 2021a).

The memory paradigm stores a subset of previously seen data from prior tasks, either directly or compressed via a generative model (Isele and Cosgun, 2018; Shin *et al.*, 2017), and utilizes a multi-task objective that combines prior tasks and the new task by replaying stored data from prior tasks during training with the new task.

The functional regularization paradigm penalizes deviation in the output of the model trained on some prior tasks when being trained on a new task, e.g., learning without forgetting (Li and Hoiem, 2016) and adaptation by distillation (Hou *et al.*, 2018).

The parameter regularization paradigm devises a measure of parameter importance for prior tasks that can be used to adaptively adjust or penalize their perturbation during training on a new task, e.g., elastic weight consolidation (Kirkpatrick *et al.*, 2017), synaptic intelligence (Zenke *et al.*, 2017) and uncertainty-guided Bayesian neural networks (Ebrahimi *et al.*, 2019).

Other paradigms can also be used for continual learning, such as Bayesian neural networks (Nguyen *et al.*, 2018), meta-learning (Al-Shedivat *et al.*, 2018) and adversarial learning (Ebrahimi *et al.*, 2020b).

These paradigms are not necessarily incompatible with each other and can be employed together to achieve the continual learning objective, such as in gradient episodic memory (Lopez-Paz and Ranzato, 2017), which employs a hybrid memory-regularization paradigm, and progress & compress (Schwarz et al., 2018), which employs a hybrid architectural-regularization paradigm.

This chapter focuses on architectural paradigms for continual learning in neural networks and, in particular, expanding neural networks. Section 2.2 discusses progressive neural networks (Rusu et al., 2016b). Section 2.3 reviews continual learning via neural pruning (Golkar et al., 2019). Section 2.4 reviews progress & compress (Schwarz et al., 2018). Section 2.5 reviews reinforced continual learning (Xu and Zhu, 2018a). Section 2.6 presents progressive learning (Fayek et al., 2020). Section 2.7 concludes the chapter.

2.2 Expanding Neural Networks

Progressive neural networks (Rusu et al., 2016b) are explicitly designed for learning tasks sequentially by employing a neural network for each task that, in addition to its own layer-to-layer connections, is connected to prior neural networks via adaptive connections, known as lateral connections, where only the latest neural network is learned using data for the new task at hand while prior neural networks are fixed, sidestepping catastrophic forgetting.

Progressive neural networks start with a standard, multi-layered neural network trained for some arbitrary task. For every subsequent task, an additional multi-layered neural network, denoted column, connected to the prior column(s) via layer-wise lateral connections in addition to its own layer-to-layer connections is added to the model and trained for the new task at hand. In other words, a fully connected layer in a column draws connections from the previous layer in the same column as well as the previous layer in the prior column(s) as follows:

$$\hat{\mathbf{y}}_{(k)}^{(l)} = \phi \left(\mathbf{W}_{(k)}^{(l)} \hat{\mathbf{y}}_{(k)}^{(l-1)} + \sum_{j<k} \mathbf{U}_{(j:k)}^{(l)} \hat{\mathbf{y}}_{(j)}^{(l-1)} \right), \qquad (2.1)$$

where $\hat{\mathbf{y}}_{(k)}^{(l)} \in \mathbb{R}^{n_{l_k}}$ is the vector of activations of layer l in column k, n_{l_k} is the number of units in layer l in column k, ϕ is an element-wise nonlinear function, such as the rectified linear unit, $\mathbf{W}_{(k)}^{(l)} \in \mathbb{R}^{n_{l_k} \times n_{(l-1)_k}}$ is the weight matrix of layer l in column k,

$\hat{\mathbf{y}}_{(k)}^{(l-1)} \in \mathbb{R}^{n_{(l-1)_k}}$ is the vector of activations of the previous layer $(l-1)$ in column k, $\mathbf{U}_{(j:k)}^{(l)} \in \mathbb{R}^{n_{l_k} \times n_{(l-1)_j}}$ are the lateral connections from layer $(l-1)$ in column j to layer l in column k and $\mathbf{y}_{(j)}^{(l-1)} \in \mathbb{R}^{n_{(l-1)_j}}$ is the vector of activations of the previous layer $(l-1)$ in column j. Only the latest column and lateral connections to prior column(s) are trained using data for the new task at hand, whereas all prior columns(s) remain constant, thus completely avoiding catastrophic forgetting.

Figure 2.1 is a procedural illustration of progressive neural networks for three tasks.

Progressive neural networks have been applied to a wide variety of tasks, including reinforcement learning (Rusu et al., 2016b), emotion recognition (Gideon et al., 2017) and frame semantic parsing (Shen et al., 2018), outperforming baselines based on independent models, fine-tuning, and other continual learning methods.

An evident limitation of progressive neural networks is the growth in the number of parameters with the number of tasks. For every new task, progressive neural networks add a new column and connections to all existing columns for prior tasks, amounting to a roughly quadratic growth in the number of parameters as a function of the number of tasks, which can be prohibitive for a large number of tasks. Thus, the focus of subsequent work has been to devise efficient architectural approaches for continual learning.

2.3 Pruning Expanding Neural Networks

An efficient way to counteract the growth in the number of parameters for expansion methods, such as progressive neural networks, is via pruning (Janowsky, 1989). Modern deep neural networks are overparameterized (Neyshabur et al., 2018), and expansion methods, such as progressive neural networks, may aggravate overparameterization due to the large number of parameters added to the model for every task.

Continual learning via neural pruning (CLNP) (Golkar et al., 2019) exploits this property by pruning the unused parameters of the neural network. Once a model is learned from the data for a task, CLNP segregates the units of the neural network into active and inactive ones. The average activity over the entire dataset is computed

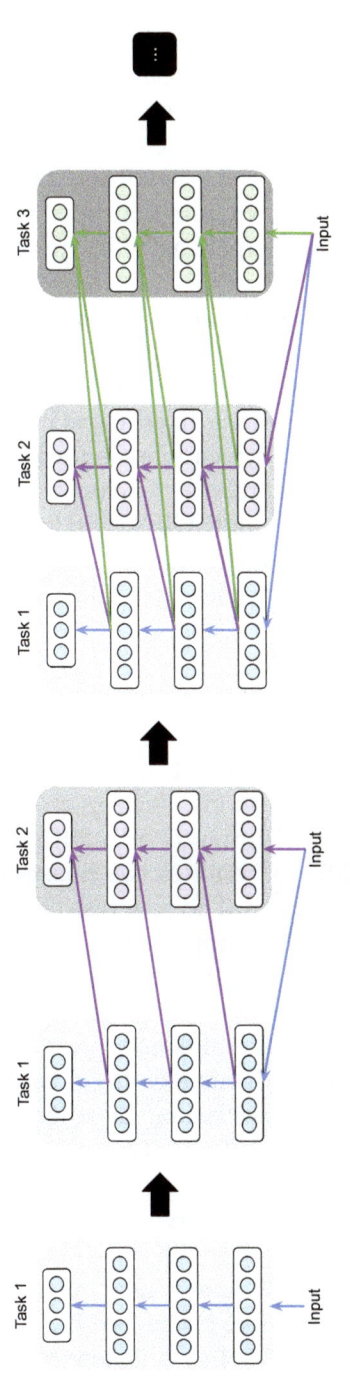

Architectural Approaches to Continual Learning 13

Fig. 2.1: Progressive neural networks for three tasks.

for each individual unit. Then, the units whose average activation exceeds some threshold value are classified as active, and those below the threshold are classified as inactive and pruned. When a new task is presented to the model, the pruned parameters are reinitialized and learned from the data for the new task. Only pruned reinitialized parameters are adapted, whereas unpruned parameters are fixed, thus avoiding catastrophic forgetting. This process is repeated for every new task.

CLNP can be viewed as adding new parameters for every new task, where the number of newly added parameters is equal to the number of pruned parameters from the prior task. Hence, CLNP is an architectural approach with a fixed capacity, unlike capacity in progressive neural networks, which increases with every task.

CLNP was applied to standard benchmark tasks, such as permuted MNIST, CIFAR-10 and CIFAR-100 (Krizhevsky, 2009), demonstrating competitive performance (Golkar *et al.*, 2019) compared with baselines (Kirkpatrick *et al.*, 2017; Zenke *et al.*, 2017).

2.4 Compressing Expanding Neural Networks

Another efficient way to counteract the growth in the number of parameters for expansion methods, such as progressive neural networks, is via distillation (Hinton *et al.*, 2015). Progress & compress (Schwarz *et al.*, 2018) alternates two phases to learn new tasks in a continual learning manner. The progress phase is similar to progressive neural networks where a new column, denoted as the active column, is added, which is a multi-layered neural network that, in addition to its own layer-to-layer connections, is connected to the existing neural network via lateral connections. The compress phase merges the active column with the existing neural network, denoted as the knowledge base, via distillation. In other words, at any given point, there are at most two neural networks, a knowledge base and an active column, where the active column is added for a new task in the progress phase and then merged into the knowledge base in the compress phase.

The progress phase introduces a new column that is a neural network composed of progressive layers, see Eq. (2.1), similar to progressive neural networks. Note that this new column is merged into the knowledge base in the subsequent phase, and thus, there is at most one column in the model in addition to the knowledge base neural network.

The compress phase consolidates the active column into the knowledge base by optimizing the following function with respect to the parameters of the knowledge base $\boldsymbol{\theta}^{\mathrm{KB}}$ while keeping the parameters of the active column unchanged:

$$\mathbb{E}\big[\mathrm{KL}(f_k(\mathbf{x})\|f^{\mathrm{KB}}(\mathbf{x}))\big] + \frac{1}{2}\|\boldsymbol{\theta}^{\mathrm{KB}} - \boldsymbol{\theta}^{\mathrm{KB}}_{k-1}\|^2_{\gamma F^*_{k-1}}, \qquad (2.2)$$

where $\mathbb{E}$ denotes the expectation over the dataset under the active column, $KL(f_k(\mathbf{x})\|f^{\mathrm{KB}}(\mathbf{x}))$ denotes the Kullback–Leibler divergence between $f_k(\mathbf{x})$ and $f^{\mathrm{KB}}(\mathbf{x})$ (Kullback and Leibler, 1951), $f_k(\mathbf{x})$ and $f^{\mathrm{KB}}(\mathbf{x})$ are the predictions of the active column after learning task k and knowledge base, respectively, given the input $\mathbf{x}$, $\boldsymbol{\theta}^{\mathrm{KB}}_{k-1}$ and F^*_{k-1} are, respectively, the mean and diagonal Fisher information matrix of the online Gaussian approximation of elastic weight consolidation (EWC) resulting from previous tasks (Schwarz et al., 2018) and γ is a hyperparameter. Note that f_k is constant throughout the compress phase.

Progress & compress unifies progressive neural networks and EWC into a single framework, capitalizing on the strengths of both methods. Progress & compress was applied to standard benchmark tasks, including reinforcement learning, and compared with baselines, including progressive neural networks (Rusu et al., 2016b) and EWC (Kirkpatrick et al., 2017), as well as other continual learning baselines (Li and Hoiem, 2016), demonstrating superior performance (Schwarz et al., 2018).

2.5 Dynamically Expanding Neural Networks

An efficient way to approach expanding neural networks for continual learning is to determine the number of parameters to add to the model, as opposed to adding a constant number of parameters for every task and then dealing with the growth in the number of parameters via pruning, distillation or some other method. Reinforced continual learning (Xu and Zhu, 2018a) adaptively expands a neural network by formulating the problem of determining the optimal number of parameters to add to the model for each task as a combinatorial optimization problem and employing reinforcement learning to solve this problem.

Reinforced continual learning is composed of three components: a controller network, a value network and a task network, all of which are implemented as neural networks, as follows.

The controller network is a long short-term memory (LSTM) recurrent neural network that determines the number of units to add for each layer given the new task at hand. The value network is a fully connected neural network that approximates the value of the state of the task network. The task network is the main neural network for learning and executing tasks.

The controller network is trained to optimize the following reward function that takes the accuracy and complexity of the expanded network into account:

$$R_k = A_k(S_k, a_{1:L}) + \alpha C_k(S_k, a_{1:L}), \qquad (2.3)$$

where A_k denotes the accuracy of the validation set of tasks k, $C_k = \sum_{l=1}^{L} w_l$ denotes the complexity of the model as the sum of the number of filters w_l added to layers $l \in \{1, \ldots, L\}$. S_k is the current state of the model, $a_{1:L}$ is a sequence of actions for L layers and α is a parameter to balance both terms in Eq. (2.3). The reward function R_k is non-differentiable; therefore, policy gradient, and in particular, actor-critic, is used to update the controller network.

Reinforced continual learning was applied to standard continual learning benchmark tasks and compared with baselines, including progressive neural networks (Rusu *et al.*, 2016b) and dynamically expandable networks (Yoon *et al.*, 2018), as well as other continual learning baselines (Kirkpatrick *et al.*, 2017; Lopez-Paz and Ranzato, 2017), demonstrating better performance in terms of accuracy as well as complexity (Xu and Zhu, 2018a).

2.6 Progressive Learning

Progressive learning (Fayek *et al.*, 2020) is a deep learning framework for continual learning. Progressive learning formulates continual learning into three procedures: curriculum, progression and pruning, as follows. The curriculum procedure selects a task to learn from

Algorithm 2.1 Functional decomposition of progressive learning in the supervised learning case (Fayek *et al.*, 2020).

Require: Candidate Tasks $\mathcal{T}_N = \{t_i \mid 1 \leq i \leq N\}$
Require: Datasets $\mathcal{D}_N = \{\mathbb{D}_i = (\mathbf{X}_{(i)}, \mathbf{Y}_{(i)}, t_i) \mid 1 \leq i \leq N\}$
Output: Model $\mathcal{F}_K = \{f_i \mid 1 \leq i \leq K\}$

1: Initialize:
$\quad \mathcal{T}_K = \varnothing$
$\quad \mathcal{D}_K = \varnothing$
$\quad \mathcal{T}_P \leftarrow \mathcal{T}_N - \mathcal{T}_K$
$\quad \mathcal{D}_P \leftarrow \mathcal{D}_N - \mathcal{D}_K$
$\quad \mathcal{F}_K = \varnothing$
2: **for** k = 1 to N **do**
3: $\quad$ *Curriculum*: select next task $t_k \in \mathcal{T}_p$
4: $\quad$ *Progression*: train f_k on task t_k using $\mathbb{D}_k = (\mathbf{X}_{(k)}, \mathbf{Y}_{(k)}, t_k)$
5: $\quad$ **if** k > 1 **then**
6: $\quad\quad$ *Pruning*: prune f_k
7: $\quad$ **end if**
8: $\quad \mathcal{T}_P \leftarrow \mathcal{T}_P - t_k$
9: $\quad \mathcal{D}_P \leftarrow \mathcal{D}_P - \mathbb{D}_k$
10: $\quad \mathcal{T}_K \leftarrow \mathcal{T}_K + t_k$
11: $\quad \mathcal{D}_K \leftarrow \mathcal{D}_K + \mathbb{D}_k$
12: $\quad \mathcal{F}_K \leftarrow \mathcal{F}_K + f_k$
13: **end for**
14: **return** Model $\mathcal{F}_K$

a set of available tasks. The progression procedure grows the capacity of the model by adding new adaptive parameters that are connected to and utilize parameters learned in prior tasks while learning from data available for the new task at hand, without being susceptible to catastrophic forgetting, similar to progressive neural networks, by fixing parameters from prior tasks. The pruning procedure counteracts the growth in the number of parameters in the progression procedure by pruning unimportant parameters.

Let $\mathcal{T}_N$ be a dynamic set of tasks, such that $\mathcal{T}_N = \mathcal{T}_K \bigcup \mathcal{T}_P$, where $\mathcal{T}_K$ is a set of learned tasks and $\mathcal{T}_P$ is a set of candidate tasks to be learned; let $\mathcal{D}_N = \{\mathbb{D}_i = (\mathbf{X}_{(i)}, \mathbf{Y}_{(i)}, t_i) \mid 1 \leq i \leq N\}$ be a set of datasets, where each dataset is associated with a task; and let $\mathcal{F}_K$ be a model of K progressive blocks, where each f_k is trained for its respective task $t_k \in \mathcal{T}_N$ using $\mathbb{D}_k \in \mathcal{D}_N$. Algorithm 2.1 is a functional decomposition of progressive learning in the supervised learning case, where the objective is to learn a model $\mathbf{y}_{(k)} = \mathcal{F}_k(\mathbf{x}_{(k)})$ for task t_k that has a dataset $\mathbb{D}_k^{(\text{train})}$ composed of M_k exemplars $\mathbf{X}_{(k)} \in \mathbb{R}^{M_k \times n_{i_k}}$ and corresponding labels or targets $\mathbf{Y}_{(k)} \in \mathbb{R}^{M_k \times n_{c_k}}$ such that n_{i_k} and n_{c_k} are the dimensionalities of the input and output, respectively, $\mathbf{x}_{(k)} \in \mathbf{X}_{(k)}$ and $\mathbf{y}_{(k)} \in \mathbf{Y}_{(k)}$. Figure 2.2 is a procedural illustration of progressive learning.

The curriculum, progression and pruning procedures can use any valid curriculum, expansion or pruning methods, respectively. These procedures can be regarded as modular building blocks for the progressive learning framework that may be employed individually or in combination as necessary. Other procedures can be added to progressive learning to address further desiderata of continual learning. The following are three simple yet effective methods for the curriculum, progression and pruning procedures, respectively:

- **Curriculum:** The objective of the curriculum procedure is to determine the most appropriate subsequent task to learn from a pool of candidate tasks with respect to the state of knowledge at the time. To this end, a simple model can be trained and evaluated for each candidate task $t_j \in \mathcal{T}_P$, using data available for the respective task $\mathbb{D}_j$ and features computed from the model learned in prior tasks f_{k-1}. The performance of each of these simple models can be used to measure the relevance of the current features in f_{k-1} to each of the candidate tasks. The subsequent task t_k can be chosen following the best performing model, or, in other words, the task that can be best solved using the already learned features.
- **Progression:** The objective of the progression procedure is to increase the capacity of the model by adding new parameters to accommodate learning a new task and, more importantly, to provide a mechanism to utilize parameters learned in prior tasks while learning the new task at hand from the data available for the task without being susceptible to catastrophic forgetting. This can be achieved by instantiating and training a new multi-layered neural network for each new task with randomly initialized parameters, denoted progressive blocks, that, in addition to its layer-to-layer connections, draws connections from the respective

Architectural Approaches to Continual Learning   19

Fig. 2.2: Overview of progressive learning for three tasks. A curriculum strategy is used to select a task (blue) from the pool of candidate tasks. Next, a model is trained to perform the selected task (blue), and the learned parameters remain constant thereafter. Then, the curriculum strategy is employed to select the subsequent task (purple). Afterward, new model parameters, denoted progressive blocks, which draw connections from the preceding layer in the block as well as the preceding layer in prior progressive block(s), are added and trained to perform the selected task (purple). Subsequently, after training the newly added progressive block to convergence, a pruning procedure is used to remove weights without compromising performance. Finally, the curriculum, progression and pruning procedures are repeated for the third task (green) and for all remaining task(s) thereafter (Fayek *et al.*, 2020).

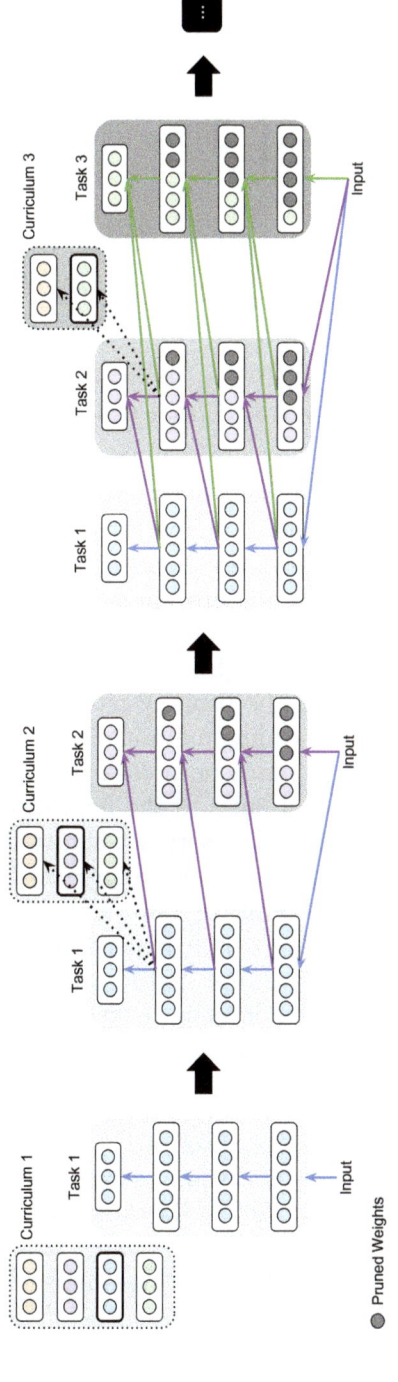

preceding layers in existing progressive blocks. A progressive, fully connected layer is defined as

$$\hat{\mathbf{y}}_{(k)}^{(l)} = \phi \left(\mathbf{W}_{(k)}^{(l)} \left(\hat{\mathbf{y}}_{(k)}^{(l-1)} \| \left(\|_{j=1}^{(k-1)} \hat{\mathbf{y}}_{(j)}^{(l-1)} \right) \right) \right), \quad (2.4)$$

where $\hat{\mathbf{y}}_{(k)}^{(l)} \in \mathbb{R}^{n_{l_k}}$ is a vector of activations of layer l in progressive block f_k, n_{l_k} is the number of units in layer l in progressive block f_k, $\mathbf{W}_{(k)}^{(l)} \in \mathbb{R}^{n_{l_k} \times \sum_{j=1}^{k} n_{(l-1)_j}}$ is a matrix of weights of layer l in progressive block f_k, $\sum_{j=1}^{k} n_{(l-1)_j}$ is the total number of units in layers $(l-1)$ in the current and prior progressive blocks, $\hat{\mathbf{y}}_{(k)}^{(l-1)} \in \mathbb{R}^{n_{(l-1)_k}}$ and $\hat{\mathbf{y}}_{(j)}^{(l-1)} \in \mathbb{R}^{n_{(l-1)_j}}$ are vectors of activations of the preceding layer $(l-1)$ in progressive blocks f_k and f_j, respectively, $l \in \{2, \ldots, L_k\}$, L_k is the total number of layers in progressive block f_k, and $\|$ denotes the concatenation operation.

- **Pruning:** The objective of the pruning procedure is to counteract the growth in the number of parameters in the model as the number of tasks increases by removing weights in a progressive block after each progression procedure. The pruning method employed is an iterative, greedy layer-wise pruning method. Following training a progressive block to convergence on a task, for each layer iteratively, starting from the initial layer to the final layer in the block f_k, the smallest q-percentile of the sorted magnitudes of the weights of the layer are removed, and the entire progressive block continues to train for a small number of iterations to compensate for the pruned weights, where q is a hyperparameter. The pruning and training procedures are repeated for an increasingly larger q, until the entire progressive layer is removed, $q \to 100\%$. The largest q that does not lead to degradation in performance is used.

Progressive learning was evaluated on a set of 11 supervised classification tasks in the image recognition domain using the CIFAR-10 and CIFAR-100 datasets (Krizhevsky, 2009). The first task was constructed from the CIFAR-10 dataset, where a training set of 45,000 images was drawn from the original CIFAR-10 training set, and the remaining 5,000 images were held out as a validation set. The test set for the first task was the 10,000 original CIFAR-10 test set images. The CIFAR-100 was used to construct the remaining 10 classification tasks, each of which was composed of 10 classes drawn from the CIFAR-100 100 classes. For each task, the training and validation sets were composed of 4,500 and 500 images, respectively, both of which were drawn from the original CIFAR-100 training set. All 1,000 images of the selected 10 classes for the task from the original CIFAR-100 test set were used as the test set for the task. The mean and standard deviation of the images were normalized to zero and one, respectively, per color channel using the training set statistics for each task individually.

The model employed was a convolutional neural network composed of four convolutional layers, followed by two fully connected layers. The first and second convolutional layers were composed of 32 filters of size 3×3, followed by batch normalization and rectified linear units. The second convolutional layer was followed by a max pooling layer with a 2×2 window and a dropout layer with a probability of 0.25. The third and fourth convolutional layers were composed of 64 filters of size 3×3, followed by batch normalization and rectified linear units. The fourth convolutional layer was followed by a max pooling layer with a 2×2 window and a dropout layer with a probability of 0.25. The first fully connected layer was composed of 512 units, followed by batch normalization, rectified linear units, and a dropout layer with a probability of 0.5. The second fully connected and final layer was followed by a softmax function.

ADAM was used to optimize the parameters with respect to a negative log-likelihood loss function using mini-batches of size 256 and a learning rate of 0.001 for 90 epochs. All layers were regularized using weight decay with a penalty of 0.001.

For progressive learning, the first task had the same architecture and settings described above. An additional progressive block was then added for each subsequent task. Each progressive block had the same architecture with half the number of filters/units of the first block in each layer. The outputs of the layer preceding a convolutional or fully connected layer in all prior blocks were concatenated and fed as input to that convolutional or fully connected layer. In each progression procedure, the parameters of the new progressive block were randomly initialized and trained using the same training recipe as described above. Note that the parameters in a progressive block

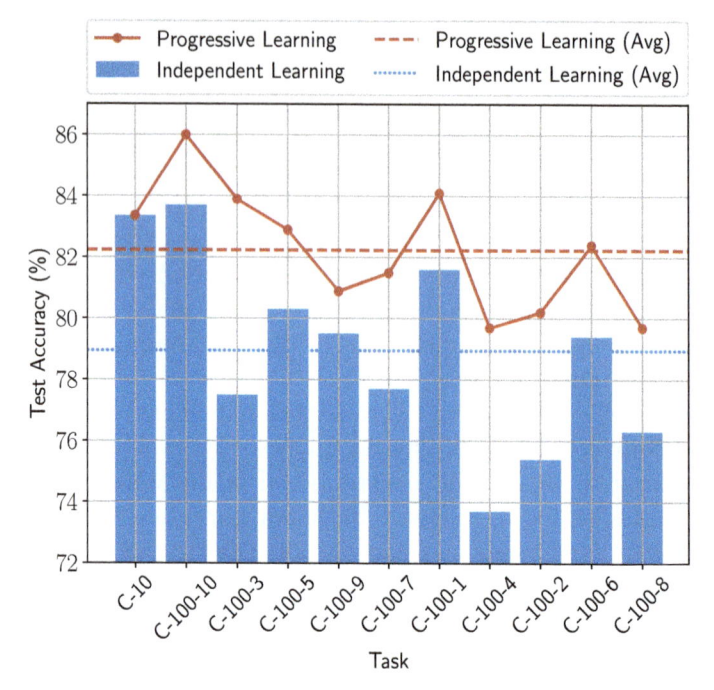

Fig. 2.3: Accuracy of progressive learning vs. independent learning for all 11 image recognition tasks. The tasks are ordered according to the outcome of the curriculum procedure. C-10 denotes the CIFAR-10 task, and C-100-s indicates the sth CIFAR-100 task (Fayek *et al.*, 2020).

are held constant once the block is trained and pruned, including during training and pruning subsequent progressive blocks.

Figure 2.3 presents the accuracy of the models for each of the 11 tasks on the test set, where it can be seen that progressive learning leads to better performance compared with independent learning by a large margin.

2.7 Discussion

The architectural approach to continual learning in neural networks is a powerful paradigm. This paradigm can not only attenuate catastrophic forgetting but also potentially completely eliminate the problem. Of course, this comes at the expense of increasing complexity. Despite recent advances in the architectural paradigm discussed above that aim to alleviate increasing complexity, other paradigms may offer more efficient solutions but are more susceptible to catastrophic forgetting. The architectural paradigm of continual learning remains a core component in state-of-the-art continual learning systems.

© 2024 World Scientific Publishing Company
https://doi.org/10.1142/9789811286711_0003

Chapter 3

Growing RBM on the Fly for Unsupervised Representation toward Classification and Regression

Savitha Ramasamy[*,†], ArulMurugan Ambikapathi[*,‡] and Kanagasabai Rajaraman[*,§]

*Institute for Infocomm Research, Agency for Science, Technology and Research (A*STAR), Singapore
†ramasamysa@i2r.a-star.edu.sg
‡a.arulmurugan@gmail.com
§kanagasa@i2r.a-star.edu.sg

Abstract

In this work, we endeavor to investigate and propose a novel unsupervised online learning algorithm, namely the online restricted Boltzmann machine (O-RBM). The O-RBM is able to construct and adapt the architecture of a restricted Boltzmann machine (RBM) artificial neural network according to the statistics of the streaming input data. Specifically, for training data that are not fully available at the onset of training, the proposed O-RBM begins with a single neuron in the hidden layer of the RBM and progressively adds and suitably adapts the network to account for the variations in streaming data distributions. Such unsupervised learning helps to effectively model the probability distribution of the entire data stream and generate robust features. We demonstrate that such unsupervised representations can be used for discriminative classifications on a set of multi-category and binary classification problems for unstructured image and structured signal data sets with varying degrees of class imbalance. We first demonstrate the O-RBM algorithm and characterize the network evolution using the simple and conventional multi-class MNIST image data set, aimed at recognizing handwritten digits. We then benchmark the O-RBM's performance against that of other machine learning, neural network and ClassRBM techniques using a number of public non-stationary data sets. Finally, we study the performance of the O-RBM on a real-world problem involving the predictive maintenance of an aircraft component and ocean environment prediction using time series data sets. In all these studies, it is observed that the O-RBM converges to a stable, concise network architecture, wherein individual neurons are inherently discriminative to the class labels despite unsupervised training. It can be observed from the performance results that, on average, the O-RBM improves accuracy by 2.5–3% over conventional offline batch learning techniques while requiring at least 24–70% fewer neurons.

3.1 Introduction

Deep learning algorithms have demonstrated great capabilities for joint mapping of features for classification, regression, segmentation and other tasks. Thus, they outperform other machine learning approaches in applications ranging from image classification (Krizhevsky *et al.*, 2012) and medical diagnostics (Turner *et al.*, 2014) to credit fraud analytics (Tomczak and Zie, 2015). However, it is challenging to adaptively retrain neural networks to track changes in input data distribution, especially in non-stationary streaming data applications in which the data are not completely available *a priori*.

Online learning approaches for deep neural networks have the potential to address this challenge. Several studies have put forth online learning algorithms for training single-layer perceptron networks (Robins, 2004; Lipton *et al.*, 2015; Dietterich, 2002). Single-layer feedforward neural networks can be trained in an online fashion using stochastic gradient descent (Bouchard *et al.*, 2016) or extended Kalman filters (Subramanian *et al.*, 2014; Suresh *et al.*, 2011). However, it remains challenging to extend these successes to the task of training deep neural networks in a fully online manner. For example, online algorithms, such as those for denoising autoencoders (DAEs) (Zhou *et al.*, 2012), have been used to achieve incremental feature learning with streaming data. However, these algorithms need *a priori* training with a DAE architecture to first learn a base set of features. Further, incremental learning has been applied within a boosting convolutional neural network framework for feature augmentation, loss function updation and fine-tuned backpropagation with information accumulating in successive mini-batches (Han et al., 2016). Finally, it has also been shown that updating greedily pre-trained layer-wise restricted Boltzmann machines (RBMs) in an online manner results in the automatic learning of discriminative features for classification (Chen et al., 2015; Savitha et al., 2018; Cote and Larochelle, 2016; Savitha et al., 2020) and regression tasks.

Recently, there has been an increased interest in lifelong continual learning approaches (Lesort et al., 2017; Schwarz et al., 2018; Parisi et al., 2019), aimed at adapting the weights of a pre-trained network to learn new tasks without catastrophic forgetting. Progressive networks (Rusu et al., 2016a) incorporate prior knowledge at initialization, retain a pool of pre-trained models throughout training, and fine-tune the model with new data to learn lateral connections for extracting useful features. A neurogenesis network (Draelos et al., 2016), inspired by the neurogenesis of the human hippocampus, that enables human beings to learn continuously and adapt themselves to changing environments has also been developed. A neurogenesis network adds neurons to a deep neural network while preserving previously trained data representations. The lifelong learning ability of a growing self-organizing neural architecture equipped with recurrent neurons for processing *time-varying patterns* is studied by Parisi et al. (2017, 2018). In order to avoid catastrophic forgetting in a continual learning framework, the representations of a deep neural network are divided into long- and short-term memory units by Li et al. (2019a), and this division is optimized with a Kalman optimizer. For complete details on the literature on continual learning, one can refer to Parisi et al. (2019) and Delange et al. (2021).

While all the above-mentioned approaches enable lifelong continual learning, they require an *a priori* trained network and/or a fixed base network architecture as a precursor for incremental online updates with streaming data. Hence, methods that evolve and train a network architecture from scratch online, as the data streams in, would offer novel capabilities and thus provide a complete online learning solution. They also enable us to understand knowledge distribution across classes, thus providing the ability to understand the data through the network. Further, it has been proven by Roux and

Bengio (2008) that the addition of a hidden neuron to an RBM helps improve its modeling power, unless it has already perfected the data. Hence, it is important to find the appropriate number of neurons in the hidden layer of a network to perfectly model any data. This can be achieved by letting the model grow its network depending on the data.

In this work, we present a fully online, unsupervised learning algorithm named the online restricted Boltzmann machine (O-RBM) for an effective and efficient feature representation. At the beginning of training, there are no neurons in the hidden layer of the RBM neural network. As training data samples stream in, the ability of the network to represent the current sample is assessed using the reconstruction error for the sample from the current architecture. Based on this reconstruction error, the online adaptive learning algorithm either deletes the samples that are well represented or adds a neuron to the hidden layer to represent the sample and updates the weights for the entire set of existing neurons in the network. As the network updates are tailored to represent the distributions of the distinctive input sample features, the network is compact and inherently discriminative (Szerlip et al., 2015). The unsupervised features that are thus learned can be mapped onto a set of specific classes via any of the conventional discriminative learning methods.

We first propose the methodology of the O-RBM, providing details pertaining to the training and evolution of the network for effective unsupervised feature representations. We then demonstrate the unique abilities of the O-RBM to represent the distinctive class distributions of the feature space and to learn in a manner that is invariant to the training data sequence through a study of the standard and well-explored MNIST data set,[1] aimed at recognizing handwritten digits. The sequential invariance is much like the invariance to permutations in the training set seen with batch learning algorithms (Poggio et al., 2011). We then analyze the performance of binary classification tasks based on the O-RBM feature representation with a variety of data sets having a wide spectrum of imbalances in the data. It is critical to learn the distribution of a minority class from a highly imbalanced data set, and the O-RBM provides a premise to

---

[1] http://yann.lecun.com/exdb/mnist/.

efficiently learn the underrepresented minority class, owing to its ability to detect novelty in the data. Our results show that the O-RBM can perform better than several state-of-the-art networks with fewer network resources than batch methods. The main contributions of the study are as follows:

- a fully online learning methodology with an evolving network architecture for unsupervised feature representation;
- an adaptive learning algorithm based on the contrastive divergence (CD) approach for training sequential data;
- empirical demonstration of the discriminative ability of the neurons trained through unsupervised O-RBM using the MNIST image (handwritten numerical characters) data set;
- rigorous empirical analysis on a variety of data sets (for varying applications), which demonstrates that the performance and the neuron-to-class label associations of the O-RBM are independent of the sequence in which the training samples are presented;
- rigorous empirical demonstrations that enumerate the abilities of the O-RBM for both classification and regression tasks.

The chapter is organized as follows: Section 3.2 introduces the architecture and learning algorithm of the O-RBM. Next, in Section 3.3, we demonstrate the ability of the O-RBM for unsupervised feature representation and subsequent classification using the MNIST data set. Following this, we show the classification and regression performance of the O-RBM in Sections 3.4 and 3.5, respectively. Finally, some interesting conclusions and potential future directions are outlined in Section 3.6.

3.2 Online Learning Restricted Boltzmann Machine

In this section, we describe the evolution of the O-RBM architecture and the corresponding learning algorithm. Figure 3.1 shows a bipartite representation of the proposed evolving RBM architecture. Specifically, we denote the training data set as $\{(\mathbf{x}^1, c^1), \ldots, (\mathbf{x}^t, c^t), \ldots, (\mathbf{x}^N, c^N)\}$. Here, $\mathbf{x}^t \in \Re^m = [x_1^t, \ldots, x_j^t, \ldots, x_m^t]^T$ is an m-dimensional input of the tth sample, $c^t \in \{1, 2, \ldots, s\}$ denotes the corresponding class labels or targets among s classes, and N is the total number of samples. Note that the

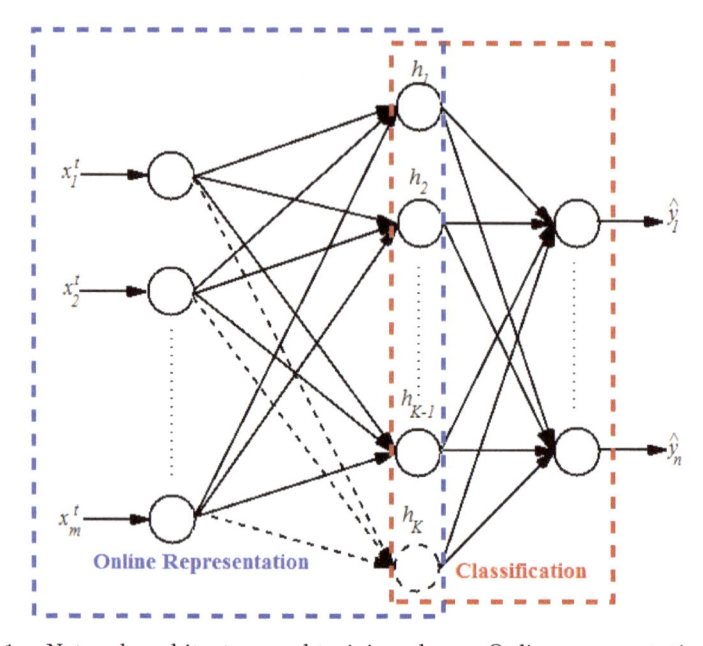

Fig. 3.1: Network architecture and training phases: Online representation learning is performed in the first phase, wherein the network begins with zero neurons in the hidden layer and adds and/or adapts as the network learning progresses to derive a feature representation of the data in an unsupervised manner. The next phase performs supervised discriminative learning to associate the feature representation with the class labels.

class labels are redundant for unsupervised feature representations. However, they can be used subsequently to evaluate the discriminative representation capability of the O-RBM. The objective of the O-RBM is to best represent the input distribution by following a fully adaptive and evolving online learning procedure.

In general, an RBM (Hinton, 2002) has visible and hidden layers, connected through symmetric weights. The number of neurons in the input layer is fixed based on the input dimension. In the current scenario, the number of neurons in the input layer is m (as $\mathbf{x} = [x_1 \cdots x_m]^T$). The hidden layer response dictates the feature representation for the input samples, and for a hidden layer with k neurons ($\mathbf{h} = [h_1 \cdots h_k]^T$), the output feature representation belongs to $\Re^k$. The feature representation is derived by learning the symmetrical connecting weights between the visible and hidden layers, i.e., w_{ij}^1; $i = 1, \ldots, m$, $j = 1, \ldots, k$, as can be inferred from Fig. 3.2.

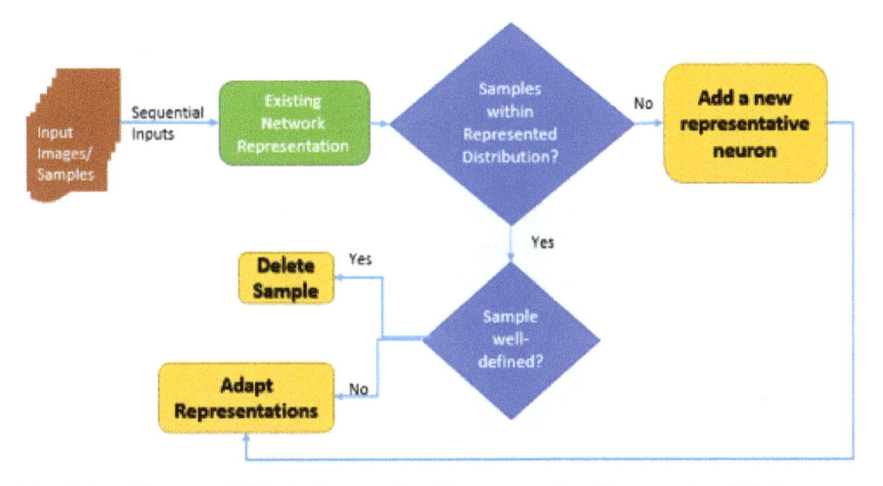

Fig. 3.2: The overall block diagram for the proposed online restricted Boltzmann machine.

The hidden layer output for the tth input sample can be expressed as

$$h_j^t = \sigma\left(\sum_{i=1}^m x_i^t w_{ij} + b_h^j\right), \quad \forall j = 1, \ldots, k, \qquad (3.1)$$

where σ is the standard sigmoid activation function and b_h^j is the bias associated with j of the k hidden neurons. It should be noted that the neurons in the same layer of the RBM are not interconnected.

3.2.1 *Evolving architecture for O-RBM*

We now describe the online learning process for feature representation at the hidden layer. Initially, the hidden layer has no neurons. As the data stream in, the online learning algorithm adds a neuron and/or updates the representations of the existing neurons, depending on the novelty of the sample. The first neuron in the hidden layer is added based on the first sample $\mathbf{x}^1$ in the training data set.

Without loss of generality, let us assume that the network comprises $k-1$ neurons in the hidden layer, corresponding to $k-1$ novel samples in a history of $t-1$ samples (with $t \gg k$). The reconstruction error of the network for the tth sample is defined as

$$E_{\text{recon}}^t = \frac{1}{m} \sum_{i=1}^{m} (x_i^t - \widehat{x}_i^t)^2, \qquad (3.2)$$

where x_i^t and $\widehat{x}_i^t$ are the ith element of the input $\mathbf{x}^t$ and the corresponding reconstructed (backward path) input element, respectively. Similar to the hidden layer representations (Eq. (3.1)), the reconstructed input can be expressed as

$$\widehat{x}_i^t = \sigma \left(\sum_{j=1}^{k} h_j^t w_{ij} + b_x^i \right), \qquad \forall i = 1, \ldots, m, \qquad (3.3)$$

where b_x^i is the bias associated with the ith input neuron.

The O-RBM algorithm chooses one of the following actions for the sample t, assessing the reconstruction error (E_{recon}^t) with respect to two predefined thresholds, namely, the novelty threshold (E_n) and the marginal representation threshold (E_m):

- **Add a Representative Neuron:** If ($E_{\text{recon}}^t > E_n$), the sample is deemed novel, and a kth neuron is added to the hidden layer of the network. The initial input weights connecting the kth hidden neuron and the neurons in the input layer are obtained as a function of the inputs $g(\mathbf{x}^t)$. Here, $g(.)$ can be any function such that $\mathbf{w}^k = [w_1 \cdots w_m]^T \in \Re^m$, and each element of $\mathbf{w}^k$ belongs to $[0, 1)$ so as to confine it within the operating regions of the network. In this work, as the inputs are normalized in $[0,1)$, we initialize $\mathbf{w}^k = 0.01 * \mathbf{x}^t$ to ensure that the weights are initialized within $[0,1)$. The network weights of all the neurons, including the new neuron, are then collectively updated according to the algorithm discussed next in Section 3.2.2.
- **Adapt the Existing Network:** If $E_n > E_{\text{recon}}^t > E_m$, the network weight matrix is adapted (as detailed in Section 3.2.2) such that the probability distribution approximated by the hidden neurons includes the representation of the current training sample.
- **Ignore Sample:** If $E_{\text{recon}}^t < E_m$, then the current tth sample is sufficiently represented by the existing network and does not warrant a network or weight matrix update.

Overall, the O-RBM architecture ensures that the neurons in the hidden layer of the network are adaptively added and updated to obtain a compact network structure that is sufficient to yield a strong feature representation for the given set of data. Having discussed the evolving strategy followed by the O-RBM, we now proceed to discuss its online learning algorithm.

3.2.2 Online CD algorithm for O-RBM

The training algorithm for the O-RBM is a subtle variation of the CD approach proposed for a conventional RBM (Hinton, 2002). For sake of completeness and clarity, the inherent details of the algorithm for the O-RBM are discussed here. Assuming k as the number of hidden neurons at the current instance of training, the energy function $E(\mathbf{x}, \mathbf{h})$ can be defined as

$$E(\mathbf{x}, \mathbf{h}) = -\mathbf{x}^T \mathbf{W} \mathbf{h} - \mathbf{x}^T \mathbf{b_x} - \mathbf{h}^T \mathbf{b_h}, \qquad (3.4)$$

where $\mathbf{x} \in \Re^m$ is the current training sample (with superscript removed for ease of representation), $\mathbf{W} \in \Re^{m \times k}$ is the weight matrix between the m inputs and k neurons, $\mathbf{h} \in \Re^k$ is the response of the k hidden neurons, and $\mathbf{b_x} \in \Re^m$ and $\mathbf{b_h} \in \Re^k$ are the biases for the m input neurons and the k hidden neurons, respectively. The probability distribution corresponding to the energy function in Eq. (3.4) can be expressed as

$$P(\mathbf{x}, \mathbf{h}) = \frac{\exp(-E(\mathbf{x}, \mathbf{h}))}{Z}, \qquad (3.5)$$

where Z is the intractable partition function that is obtained by averaging $\exp(-E(\mathbf{x}, \mathbf{h}))$ over all possible values of $\mathbf{x}$ and $\mathbf{h}$ so that the probability distribution $P(\mathbf{x}, \mathbf{h})$ sums to one. Subsequently, the marginal probability of the input $\mathbf{x}$ vector can be obtained from $P(\mathbf{x}, \mathbf{h})$ by summing over all possible hidden layer configurations, i.e.,

$$P(\mathbf{x}) = \frac{1}{Z} \sum_{\mathbf{h}} \exp(-E(\mathbf{x}, \mathbf{h})). \qquad (3.6)$$

The purpose of a training algorithm is then to maximize the expectation of $\log P(\mathbf{x})$, for any given $\mathbf{x}$. However, maximization over the probability function involves the computation of expectations over the entire distribution. Hence, the CD approach, which approximates the original distribution with a sample point estimation, is being employed. However, unlike conventional backpropagation, CD employs backprojection (reconstruction) and the Gibbs sampling procedure to estimate the weights. Mathematically, CD considers maximizing the following optimization problem:

$$\arg\max_W E[\log P(\mathbf{x})]. \tag{3.7}$$

Then, for a training sample $P(\mathbf{x})$, the CD algorithm aims to compute the conditional probabilities of the hidden units as follows (as the hidden units are not interconnected and hence independent of each other):

$$P(\mathbf{h}|\mathbf{x}) = \prod_{i=1}^{K} P(h_i|\mathbf{x}), \tag{3.8}$$

where the individual conditional probabilities $P(h_i|\mathbf{x})$ are computed using sigmoidal activations (Eq. (3.1)). We then sample a hidden feature vector $\mathbf{h}$ based on the obtained distribution. Using the current hidden feature vector $\mathbf{h}$, a reconstruction of $\mathbf{x}$, namely $\widehat{\mathbf{x}}$ (Eq. (3.3)), is obtained by sampling the following distribution (as the input nodes are not interconnected and hence independent of each other):

$$P(\widehat{\mathbf{x}}|\mathbf{h}) = \prod_{i=1}^{m} P(\hat{x}_i|\mathbf{h}). \tag{3.9}$$

Again, we compute the hidden nodes' conditional probabilities $P(\mathbf{h}|\widehat{\mathbf{x}})$ based on the reconstructed $\widehat{\mathbf{x}}$ by following Eq. (3.8) and perform Gibbs sampling to sample a hidden feature vector $\widehat{\mathbf{h}}$. Then, the element-wise weights between the input and hidden layers of the network are updated according to

$$w_{ji} = w'_{ji} + \alpha * (x_j * h_i - \hat{x}_j * \hat{h}_i), \tag{3.10}$$

wherein α denotes a prespecified learning rate. Also, the respective biases of the input and hidden nodes can be updated as

$$\mathbf{b_x} = \alpha * (\mathbf{x} - \widehat{\mathbf{x}}), \tag{3.11}$$

$$\mathbf{b_h} = \alpha * (\mathbf{h} - \widehat{\mathbf{h}}). \tag{3.12}$$

Algorithm 3.1 Online learning of RBM.

Step 0: Given a streaming train data set, $\{\mathbf{x}^1, \ldots, \mathbf{x}^t, \ldots, \mathbf{x}^N\}$, for each sample $\mathbf{x}^i$ in the training set, do the following.

Step 1: Choose an optimal value for the novelty threshold E_n and the marginal representation threshold E_m.

Step 2: Obtain a sample hidden vector representation by sampling the distribution obtained using Eq. (3.8).

Step 3: Backproject the sampled $\mathbf{h}$ to obtain a sample of the reconstructed input distribution $\hat{\mathbf{x}}$, as given by Eq. (3.9).

Step 4: Compute the reconstruction error between $\mathbf{x}$ and $\hat{\mathbf{x}}$, as given by Eq. (3.2). If a new hidden neuron needs to be added and/or the hidden neurons need to be updated, go to **Step 4**; else, go to the next sample (**Step 1**).

Step 5: For $\hat{\mathbf{x}}$, compute the hidden vector distribution by following Eq. (3.8) and perform Gibbs sampling (once) to get $\hat{\mathbf{h}}$.

Step 6: The resultant gradient computation and the element-wise weight update can be done by following the gradient ascent given by Eq. (3.10). The corresponding bias update is given by Eq. (3.11).

Step 7: Repeat **Steps 1–6** sequentially until all the training samples are considered.

We have opted for a single-step CD approach wherein the sampling (input and hidden feature representations) is done only once (in both the visible and hidden layers) per update (Bengio and Delalleau, 2009). It should be noted that, while the dimension of $\mathbf{x}$ remains the same throughout the training, the dimension of $\mathbf{h}$ keeps varying according to the number of hidden neurons at the current instance of training. The weight and bias updates should be done for each training sample instance if the constraints stipulated in Section 3.2.1 are met. The entire online learning procedure for O-RBM training is summarized in Algorithm 3.1.

3.2.3 Discriminative training

For sake of continuity, we briefly discuss discriminative training, where the feature representation learned during the online generative phase is mapped onto the conditional class distributions in a supervised fashion.

The responses of the K neurons in the hidden layer are as follows:

$$\mathbf{h} = [h_1 \cdots h_k]^T. \tag{3.13}$$

This feature representation is then used in a supervised discriminative training phase to learn the conditional probability distribution $P(c^t|\mathbf{x}^t)$. For classification problems, the class labels c^t are encoded in $\mathbf{y}^t = [y_1^t, \ldots, y_s^t]$ as follows:

$$y_i^t = \begin{cases} 1 & \text{if } c^t = i, \\ 0 & \text{otherwise,} \end{cases} \quad i = 1, \ldots, s. \tag{3.14}$$

The objective of discriminative training is to minimize the log probability

$$\min_{\mathbf{w}_{ki}^2} \frac{1}{N} \sum_{n \in N} \mathcal{L}_{\text{disc}}\left(\mathbf{y}^n | \mathbf{x}^n\right), \tag{3.15}$$

where $\mathcal{L}_{\text{disc}}(\mathbf{y}^n|\mathbf{x}^n)$ is a measure of the error between $\mathbf{y}^n$ and $\widehat{\mathbf{y}}^{\mathbf{n}}$, and $\mathbf{w}_{ki}^2$ are the weights connecting the kth output neuron and the ith hidden neuron. Here, we perform discriminative training through 10 epochs of supervised training using a multi-layer perceptron (MLP) with a sigmoidal activation function.

3.3 Demonstration of O-RBM Learning

We now demonstrate the progression of learning with the proposed O-RBM approach and make some observations about the algorithm. We characterize the algorithm using the MNIST data set (Lecun et al., 1998), as it is a large, well-explored multi-category data set (60,000 training samples, 10 categories). The network is trained online using the training data set. The validity of the trained network is established independently on the test set (10,000 samples, 10 categories) in an offline manner.

Figures 3.3(a) and 3.3(b) show the evolution of reconstruction error and network architecture as samples stream in for training. Figure 3.3(a) shows that the reconstruction error is high for the

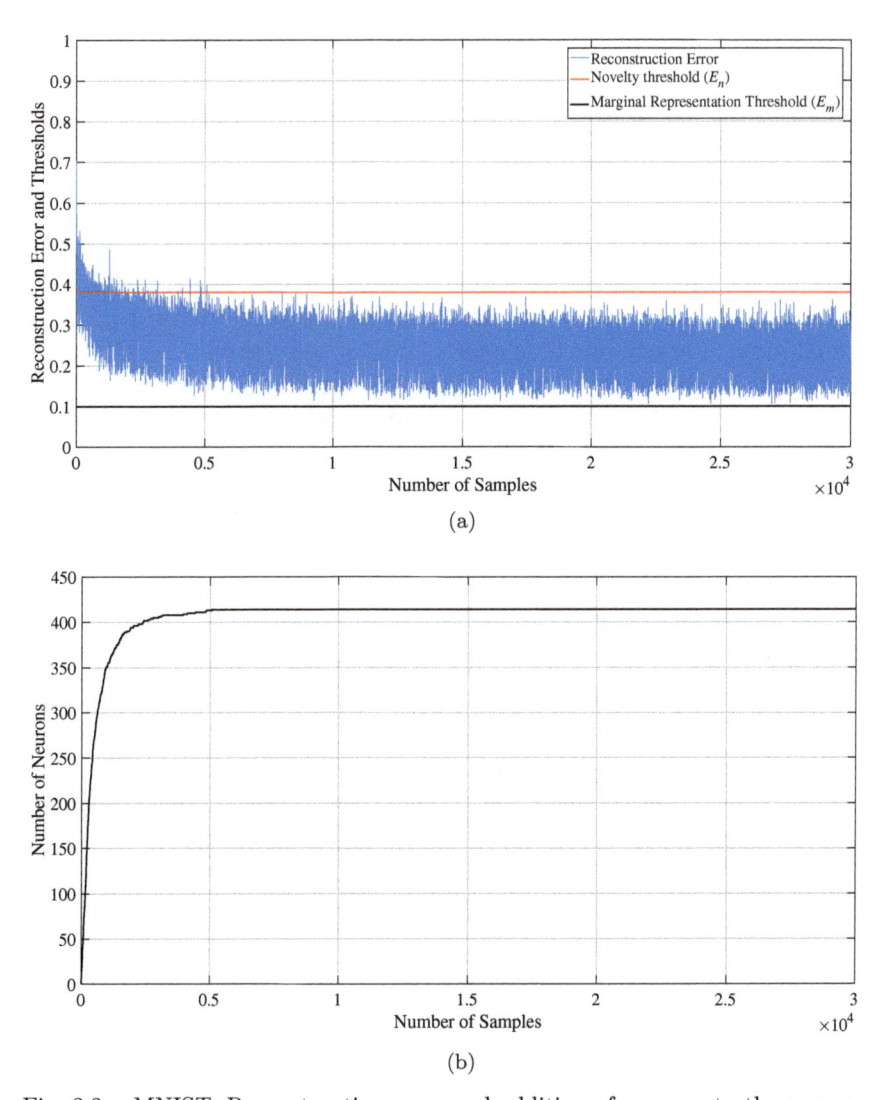

Fig. 3.3: MNIST: Reconstruction error and addition of neurons to the generative RBM network through online learning: (a) reconstruction error of samples 1–30,000; (b) growth of the network with samples 1–30,000.

initial samples. This is because the model is at its infancy and is beginning to learn. Hence, most samples are novel to the network, resulting in neurons being added (see Fig. 3.3(b)). However, as training progresses, the network learns a sufficient representation of the data, and the reconstruction error reduces progressively, resulting in fewer neurons being added to the network. It is evident from Figs. 3.3(a) and 3.3(b) that the online generative phase converges to a stable, concise network architecture, and the generative training is complete in about 26 min. It is also evident from Fig. 3.3(b) that 90% of the neurons in the stable network are added for the first 10% of the training samples (i.e., the first 5,000 samples). The remaining 90% of the training samples (i.e., the latter 55,000 samples) contribute only about 10% of the neurons in the stable network. Next, we conduct a study on the choice of the novelty threshold (E_n). We do this by training the network using the training data set, varying the novelty threshold from 0.1 to 0.5. We hold a validation data set of 1,000 samples by sampling 100 samples from each class in the testing data set. With the addition of each neuron, the reconstruction error of a subset of the testing data set is computed. It can be observed from Fig. 3.4 that the choice of E_n affects the convergence, especially because the network learns in a single pass. Setting a high threshold for E_n results in too few neurons, and vice versa, with poor learning in either case. On the other hand, setting E_n in the range of 0.25–0.3 results in optimal architecture and optimal learning.

We next validate the effect of the sequence in which the training data are presented on the performance of the algorithm. We train the O-RBM independently for 50 randomly constructed sequences

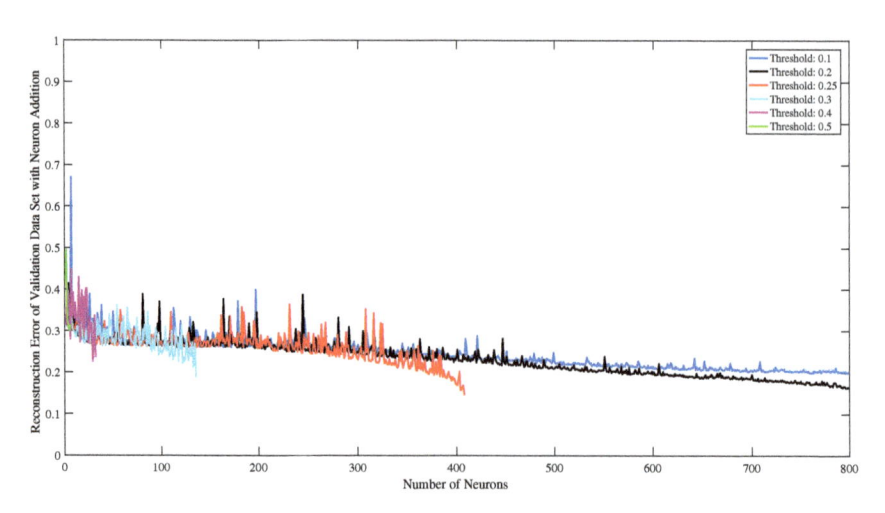

Fig. 3.4: Choice of threshold, E_n, of O-RBM for the MNIST data set.

Growing RBM on the Fly for Unsupervised Representation   39

of the MNIST training samples. In each case, we present different sequences of the training data set to train the network. Across the 50 training trials, the classification accuracy on the testing data set is $97\pm2\%$, and the final number of neurons is 403 ± 26. Thus, changing the sequence of presentation of the training samples does not change the accuracy or the network architecture significantly, showing that the network is able to generalize well with a concise network architecture.

To evaluate the discriminative potential of the feature representation learned during the online generative training phase, we relate the number of "novel" samples (where $E^t_{\text{recon}} > E_n$) to their corresponding class labels c^t for each of the 50 trials. Figure 3.5 shows the average number of hidden layer neurons associated with each class of the MNIST data set, with the standard deviation across the 50 trials. These results show that the individual neurons in the trained network are inherently and collectively discriminative toward the class labels, despite the unsupervised nature of the training. Further, we observe that the variability across trials is a small proportion of the average number of neurons in each class, suggesting that the neuron-to-class associations are largely independent of the sequence of the training data samples.

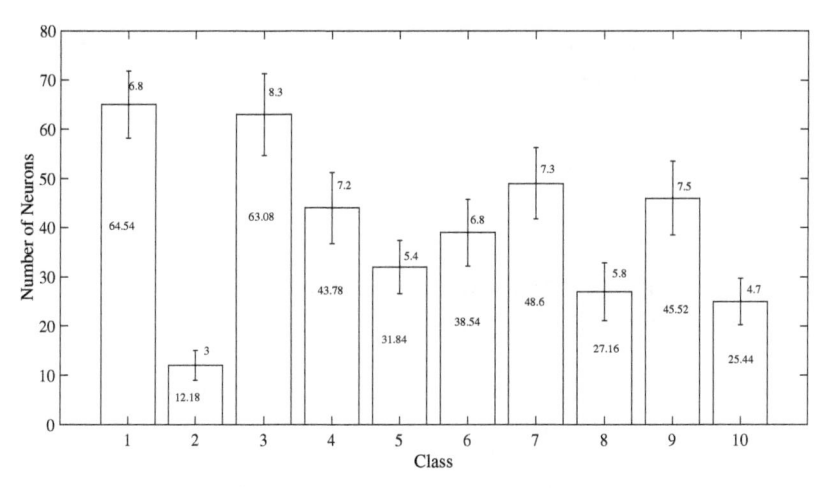

Fig. 3.5: MNIST classification: Average number of hidden layer neurons associated with each class of the MNIST data set, with the standard deviation across 50 trials.

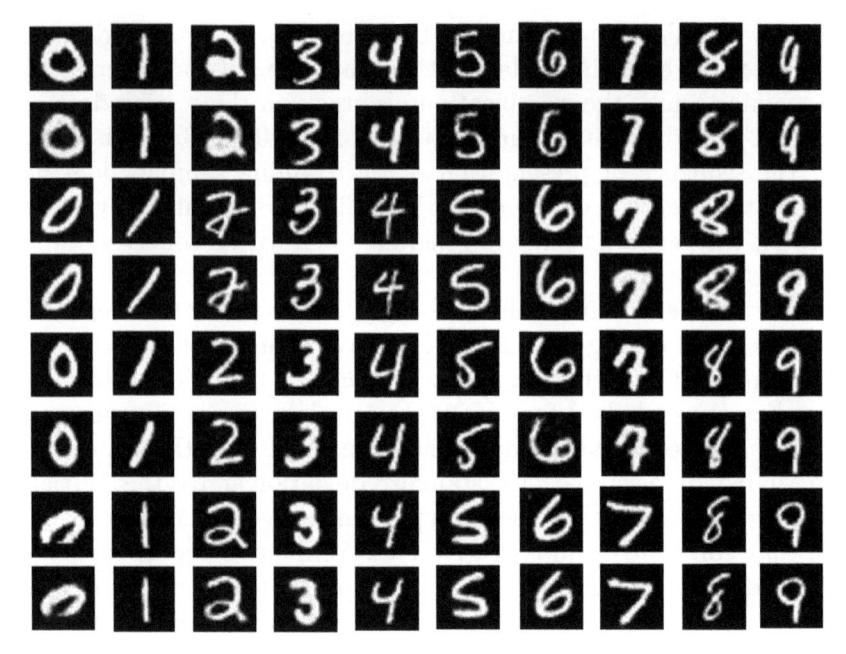

Fig. 3.6: MNIST classification: Reconstruction of the handwritten digits using the online generative network. The odd-numbered rows represent the original image, while the even-numbered rows are the images reconstructed by the RBM network trained with the O-RBM.

To demonstrate the accuracy of reconstructions, we present a subset of the reconstructed image in comparison to the original image of the MNIST data set in Fig. 3.6. It can be observed from the figure that, despite single-pass learning, the O-RBM is capable of accurate reconstructions.

3.4 Performance Study of Classification Tasks: Credit Fraud Analytics

As online learning algorithms are particularly suitable for streaming data applications with an evolving stream of data, we analyze the performance of the O-RBM with those of batch learning and online learning techniques for single-hidden-layer neural networks. Specifically, we evaluate the performance of O-RBM on the problem of credit scoring of borrowers. We compare the various classifiers on these problems based on the network size and performance measures such as the overall efficiency (η_O), average efficiency (η_A), true positive rate (TPR), true negative rate (TNR) and geometric mean accuracy (Gmean), which are defined as follows:

$$\eta_O = \frac{\sum_{i=1}^{s} q_{ii}}{N} \times 100\%, \qquad (3.16)$$

$$\eta_A = \frac{1}{s} \sum_{i=1}^{s} \frac{q_{ii}}{N_i} \times 100\%, \qquad (3.17)$$

$$\text{TPR} = \frac{\text{Number of TP}}{\text{Number of TP} + \text{Number of FN}}, \qquad (3.18)$$

$$\text{TNR} = \frac{\text{Number of TN}}{\text{Number of TN} + \text{Number of FP}}, \qquad (3.19)$$

$$\text{Gmean} = \sqrt{\text{TPR} \times \text{TNR}}. \qquad (3.20)$$

Here, q_{ii} is the number of correctly classified samples in class i and denotes the diagonal elements of the confusion matrix $Q \in \Re^{s \times s}$ of s classes. The number of samples in class i is denoted by N_i.

Credit scoring is a problem of estimating the probability that a borrower might default and/or exhibit undesirable behavior in the future. The credit patterns of individual borrowers are bound to evolve over time, with huge interpersonal variability across borrowers, and are often characterized by severe imbalances in the data, as the defaulters are few and far between. The O-RBM provides a better premise to represent such evolving time series data with severe class imbalances owing to its ability to detect novel samples better.

Several studies have employed batch machine learning techniques for credit scoring (Baesens et al., 2003; Lessmann et al., 2015; Hand and Henley, 1997; Tomczak and Zie, 2015). We perform analogous evaluations to benchmark our online learning algorithm in relation to these batch learning techniques. Specifically, we perform credit fraud prediction using three publicly available data sets, namely, the UCI German credit data set (UCI German), the UCI Australian credit data set (UCI AUS) and the Kaggle "Give me some credit" data set (Kaggle GMSC). We evaluate the O-RBM classifier in comparison with the support vector machine (SVM) classifier, the MLP neural network (NN) classifier, the classification restricted

Table 3.1: Description of the credit scoring data sets.

Data set	Input features	Size of data set	IF
UCI German	24	1,000	0.4
UCI AUS	14	690	0.1101
Kaggle GMSC	10	1,50,000	0.86632

Boltzmann machine (ClassRBM) classifier (Larochelle *et al.*, 2012) and the scoring table (ST) method on the three credit data sets (listed in Table 3.1), as shown in Table 3.2.

Table 3.1 details the public credit scoring data sets, along with the number of classes, the number of training and testing samples and their imbalance factors (IF):

$$\text{IF} = 1 - \frac{s}{N} \min_{i=1,\ldots,s} N_i, \quad (3.21)$$

where s is the total number of classes and N_i is the number of samples in class i. It is evident that the three public data sets have varying degrees of class imbalance. While UCI AUS is mildly imbalanced, UCI German is partially imbalanced and Kaggle GMSC has a very high imbalance across classes. This varying degree of class imbalance provides a unique opportunity to characterize the neuron distribution across classes in the online learning framework. We filled in the missing values in the Kaggle GMSC data set by averaging across similar participants in the population, grouped according to ages in intervals of 10.

We now present the results of the O-RBM in relation to the batch learning and state-of-the-art online learning techniques for a single-hidden-layer network in Table 3.2. For batch learning techniques, we compare the performances of the O-RBM with those of SVM, NN and ClassRBM (Tomczak and Zie, 2015). On the other hand, the online learning performance of the O-RBM is benchmarked against the projection-based learning algorithm of a radial basis function network (PBL-RBFN). We reproduce previously obtained batch learning results using the SVM, NN, ClassRBM and ST classifiers from Tomczak and Zie (2015). The ClassRBM results from Tomczak and

Table 3.2: Performance analysis of benchmark data sets: Credit scoring.

Data set	Classifier	K	Training η_O	η_A	Testing η_O	η_A	TPR	TNR	Gmean
UCI German	SVM	534	76.43	66.68	74.67	61.38	0.33	0.89	0.54
	SVM*	—	—	—	—	—	0.48	0.87	0.65
	NN	60	98.57	97.57	72.33	65.11	0.46	0.84	0.62
	NN*	—	—	—	—	—	0.52	0.81	0.65
	ClassRBM	80	77.43	63.35	74.00	56.74	0.44	0.83	0.60
	ClassRBM*	100	—	—	—	—	0.48	0.87	0.65
	ST*	—	—	—	—	—	0.67	0.68	0.68
	PBL-McRBFN	100	80.14	81.78	67.33	69.11	0.65	0.73	0.69
	O-RBM	48 (32:16)	79	74.2	76.5	71.69	0.60	0.83	0.71
UCI AUS	SVM	192	85.51	86.26	85.51	86.05	0.80	0.93	0.86
	SVM*	—	—	—	—	—	0.91	0.71	0.850
	NN	60	94.82	94.77	84.06	83.73	0.79	0.88	0.84
	NN*	—	—	—	—	—	0.85	0.86	0.85
	ClassRBM	50	86.13	86.39	85.51	86.02	0.90	0.83	0.86
	ClassRBM*	100	—	—	—	—	0.88	0.85	0.86
	ST*	—	—	—	—	—	0.83	0.81	0.82
	PBL-McRBFN	34	85.71	86.26	81.40	80.60	0.85	0.76	0.80
	O-RBM	38 (20:18)	86.68	86.8	88.49	89	0.92	0.86	0.89
Kaggle GMSC	SVM	6340	69.97	59.43	72.24	60.02	0.58	0.90	0.72
	SVM*	—	—	—	—	—	0.11	0.99	0.34
	NN	60	63.90	62.29	74.20	63.02	0.62	0.88	0.74
	NN*	—	—	—	—	—	0.23	0.99	0.48
	ClassRBM	100	75.69	74.05	86.16	74.79	0.6	0.90	0.73
	ClassRBM*	100	—	—	—	—	0.18	0.99	0.42
	ST*	—	—	—	—	—	0.52	0.62	0.57
	PBL-McRBFN	11	57.04	53.31	80.52	53.18	0.22	0.85	0.43
	O-RBM	13 (3:10)	76.08	74.49	86.25	75.22	0.63	0.88	0.74

Source: *Reproduced from Tomczak and Zie (2015).

Zie (2015) are reported with a fixed architecture of 100 neurons having a batch size of 100 and a learning rate of 0.0001, whereas the architectures of the other classifiers are not specified. Further, the training accuracies of the classifiers have also not been reported. Hence, we perform independent evaluations using the SVM, NN and ClassRBM classifiers to report additional performance validation beyond the previously reported results.

From the table, we can observe that the online learning algorithms require fewer neurons, in general, compared to the batch learning techniques. This can be attributed to the ability of online learning algorithms to evolve their network architectures depending on the novelty of the samples in the data. Further, the online learning algorithms are capable of generalizing well with better performance on the testing data, compared with the other algorithms. Thus, although the O-RBM exhibits training accuracy comparable to the batch learning techniques, it shows a better ability to generalize on a held-out test set. The ability of the O-RBM to represent the minority class is exemplary in comparison to the batch learning techniques, especially on the very highly imbalanced Kaggle GMSC data set. It is encouraging to observe that, in general, the O-RBM is better than the state-of-the-art single-hidden-layer neural network with batch learning algorithm by at least 2%.

3.5 Performance Study on Regression Tasks: Ocean Wave Height Prediction

In this section, we evaluate the ability of the proposed O-RBM for regression by reporting its ability to predict wave height in diverse geographical locations. We use data from three regions for training: the Gulf of Mexico, the Korean region and the UK region. The details of the stations from which the data are collected and their latitudes and longitudes are summarized in Table 3.3. For complete details, one may refer to Kumar *et al.* (2018). We use the data from January 1, 2011, to December 31, 2014, for training, and the data from January 1, 2015, to August 31, 2015, are used for testing.

3.5.1 *Online learning for regression of ocean wave height in independent regions*

First, we evaluate the performance of the O-RBM to predict wave height at the stations of the various regions, namely, the Gulf of Mexico and the Korean and UK regions. This will help ascertain the regression performance of the O-RBM. The representations that are obtained through unsupervised online learning of the O-RBM are used to train a neural network using a small amount of data from the independent geographical regions, such that the model is able to predict the wave height at the end of training. The neural network model training uses the representations obtained from the O-RBM,

Table 3.3: Data characteristics.

Utility of data	Region	Station ID	Latitude	Longitude
Training	Gulf of Mexico	42001	25.888	−89.658
		42002	26.091	−93.758
		42019	27.907	−95.353
		42035	29.232	−94.413
	Korean	22103	34	127.5
		22105	37.53	130
		21229	37.46	131.11
		22108	36.25	125.75
	UK	62029	48.7	−12.5
		62107	50.1	−6.1
		62170	51.4	2
		62052	48.5	−5.8
		62106	49.9	−2.9
Transfer	Irish	62091	53.48	−5.43
		62092	51.13	−10.33
		62093	55	−10
		62094	51.41	−6.42
		62095	53.04	−15.53

and the weights connecting the representations and the output layer are learned through gradient descent learning, which is based on the mean square error of the target wave height. In all these studies, the root mean square error (RMSE) of the prediction is estimated through

$$\text{RMSE} = \sqrt{\frac{1}{N}(y_y - \widehat{y}_t)^2}, \quad (3.22)$$

where N is the total number of samples in the test set, y_t is the label for the sample and $\widehat{y}_t$ is the predicted output for the sample t.

The RMSE of the predictions for the stations in the three regions of the training data are tabulated in Table 3.4. From the table, it can be observed that the network trained using the representations of the O-RBM outperforms the state-of-the-art algorithms in predicting the ocean wave height by a substantial margin.

Figure 3.7 shows a snapshot of the evolution of the novelty threshold E_n, along with the neuron growth in the hidden layer of the O-RBM, with streaming data. From the figure, it can be observed

Table 3.4: Performance study results of wave height prediction: Test data in the individual source regions.

Region	Station	SVR	ELM	GAP-RBF	MRAN	DBN	O-RBM
Gulf of Mexico	42021	0.57	0.47	0.27	0.24	0.22	**0.08**
	42002	0.49	0.22	0.21	0.20	0.19	**0.08**
	42019	0.69	0.30	0.47	0.24	0.23	**0.09**
	42035	0.75	0.19	0.22	0.21	0.16	**0.04**
Korean	22105	0.70	0.66	0.64	0.50	0.36	**0.17**
	22103	0.65	0.52	0.48	0.44	0.35	**0.15**
	21229	0.65	0.52	0.48	0.44	0.35	**0.17**
	22108	0.41	0.36	0.33	0.24	0.24	**0.13**
UK	62029	0.36	0.61	0.37	0.23	0.20	**0.77**
	62107	0.56	0.67	0.67	0.38	0.21	**0.82**
	62170	0.67	0.74	0.66	0.54	0.43	**0.47**
	62103	0.66	0.58	0.45	0.50	0.42	**0.21**

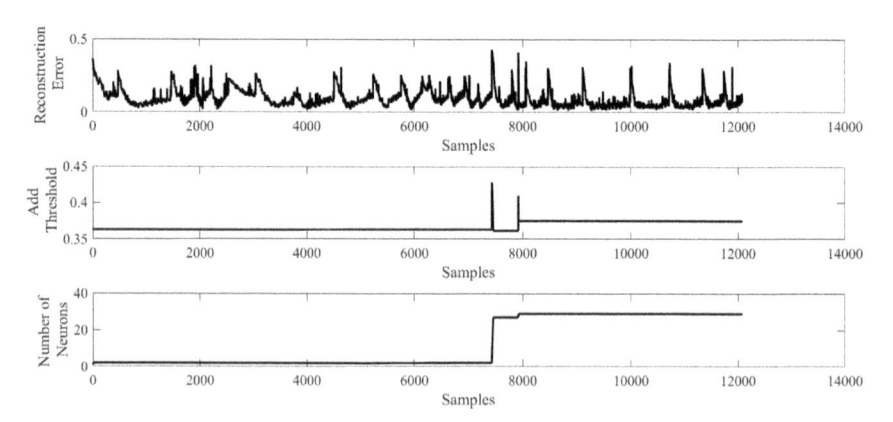

Fig. 3.7: The evolution of add threshold and the network hidden size, with streaming data in the UK region.

that the network adds neurons according to the statistics of the data as they evolve dynamically. The novelty threshold (E_n) shoots up after 7,000 samples, when the error increases substantially, and the number of neurons increases as the error increases.

3.5.2 Performance results of transferring representations of WHP prediction using online incremental learning

Finally, we present the performance of the wave height prediction model, when the training data are learned online. Accordingly, the weights between the input and hidden layers of the RBM are learned online incrementally using data from the stations in the Gulf of Mexico and in the Korean and UK regions. Thereafter, these representations are transferred to learn the output weights of the RBM using the data from the Irish region toward predicting wave heights.

Figure 3.8 presents the learning characteristics of the O-RBM for learning incrementally using data from multiple sites. The green line represents the end of training using data from stations in the Gulf of Mexico, and the blue line represents the end of training using data from stations in the Korean region. It is observed that only 5 neurons are added based on data from the stations in the Gulf of Mexico. On the other hand, as the O-RBM learns incrementally using data from the stations in the Korean and UK regions, 28 and 7 neurons are added, respectively. Thus, at the end of training using the data from the stations in all three regions, a total of 40 neurons are added to

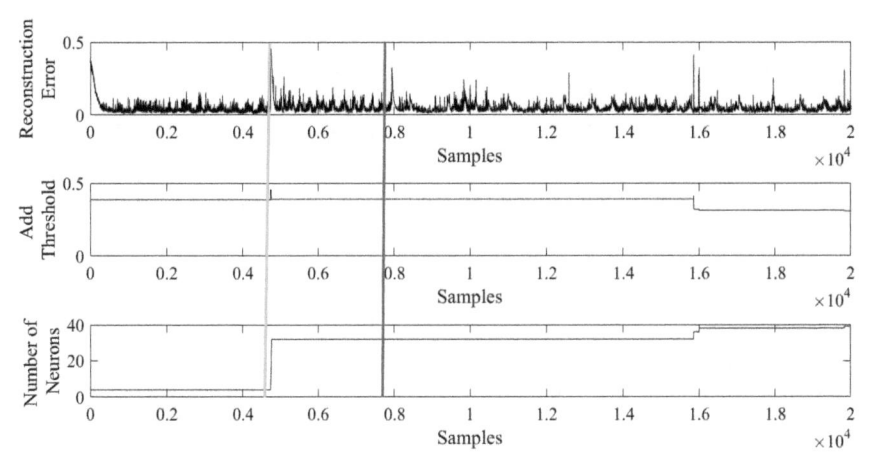

Fig. 3.8: Incremental learning of O-RBM using data from multiple regions: The green line represents the end of training using data from stations in the Gulf of Mexico, and the blue line represents the end of training using data from stations in the Korean region.

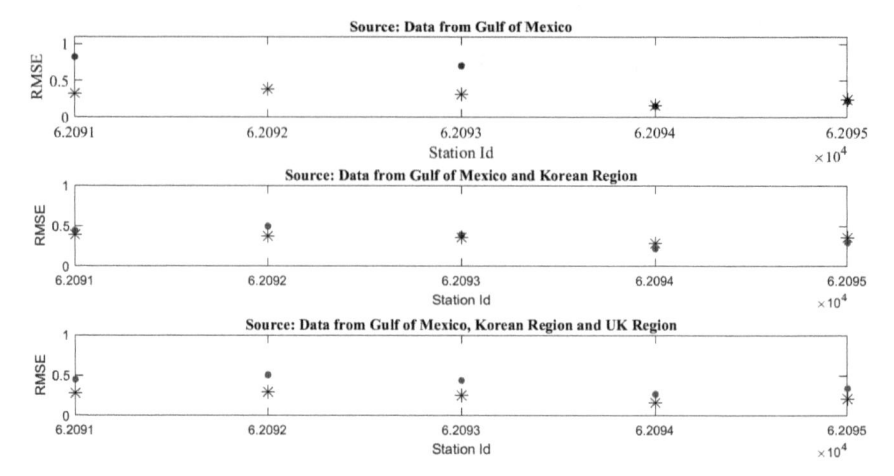

Fig. 3.9: Transferring representations from source regions to the Irish region: RMSEs of DBN vs. O-RBM.

the network. It must be noted that the proposed method does not require data to be saved and/or shared, thus ensuring the privacy of data across stations.

Figure 3.9 presents the results of transferring these learned representations to train the output weights using the data from the Irish region. The prediction performance of the model in comparison with the DBN is presented in Fig. 3.9. For a fair comparison, the DBN is also trained with a single hidden layer with 40 neurons using data from the regions, as shown. From the figure, it can be observed that the transfer of the O-RBM representations is more accurate than that of the DBN. Furthermore, it can be observed that the model that is trained incrementally on more data has better prediction accuracies.

Generally, we could make the following inferences from the performance study results on the various problems:

- **Network Size:** Overall, the O-RBM network uses fewer neurons than the other single-hidden-layer classifiers used in comparison. This is because the O-RBM uses the most novel samples to add neurons to the network, and the neurons are well representative of the data set.
- **Performance Measures:** Despite having a compact architecture, the proposed O-RBM performs better than all the other models used in comparison. This could be attributed to the fact that the

learned distributions represent the data very well. Moreover, while the other algorithms learn the data in batches and update gradients in batches, the O-RBM updates gradients based on every sample in the data set.
- **Neuron Distribution Per Class:** Unlike batch learning algorithms that need an *a priori* assumption of the architecture, the O-RBM builds the network as learning progresses. In classification problems, this helps in inferring the number of neurons per class, which may aid in characterizing the distribution of the samples in each class.
- **Effect of Class Imbalance in Classification:** Classes with fewer samples require more neurons for sufficient feature representation. As the class imbalance increases, a greater proportion of the hidden layer neurons are associated with less prevalent classes. This adaptation is a natural consequence of the online learning process and differentiates our approach from the batch learning algorithms.
- **Transferability of Representations:** With its continual learning approach, the O-RBM is capable of continual learning online, allowing it to transfer representations across data sets with diverse distributions.

3.6 Conclusion

We introduced a novel O-RBM framework that evolves a network architecture in a fully bottom-up, online manner as data streams in. We demonstrated that the algorithm converges to a stable, compact network architecture, wherein the hidden layer neurons are implicitly associated with class labels (despite unsupervised training) and classification performance is invariant to the sequence in which the training data samples are presented. Further, the O-RBM performed better than batch techniques in credit score classification and ocean wave height regression, with streaming data. Specifically, online learning achieved better accuracy with fewer neurons and showed the unique ability to adapt to class imbalances. Designing online learning frameworks for a multi-layered deep belief network is currently under investigation. We believe that this work will foster online learning research in areas that include convolutional neural network

(CNN)-based supervised and unsupervised learning and generative and interpretable models.

Acknowledgments

We acknowledge the discussions with several colleagues, including Krishna Kumar Nagalingam, Cheryl Wong Sze Yin, Gopinathan Arjunan, Guo Yang and Ponnuthurai Nagaratnam Suganthan.

© 2024 World Scientific Publishing Company
https://doi.org/10.1142/9789811286711_0004

Chapter 4

Lifelong Learning for Deep Neural Networks with Bayesian Principles

Cuong V. Nguyen[*,¶], Siddharth Swaroop[†,**], Thang D. Bui[‡,††], Yingzhen Li[§,‡‡] and Richard E. Turner[∥,§§]

[*]*Department of Mathematical Sciences, Durham University, UK*
[†]*Department of Computer Science, Harvard University, USA*
[‡]*School of Computing, Australian National University, Australia*
[§]*Department of Computing, Imperial College London, UK*
[∥]*Department of Engineering, University of Cambridge, UK*
[¶]*viet.c.nguyen@durham.ac.uk*
[**]*siddharth@seas.harvard.edu*
[††]*thang.bui@anu.edu.au*
[‡‡]*yingzhen.li@imperial.ac.uk*
[§§]*ret26@cam.ac.uk*

Abstract

This chapter describes a general Bayesian framework for the lifelong learning of artificial neural networks that can handle catastrophic forgetting in a principled way. The framework can be applied to both discriminative and generative models as well as task-aware and task-agnostic settings. We introduce the variational continual learning algorithm, a realization of this framework that uses online variational inference with a small amount of memory or coreset for effective lifelong learning. We examine various practical considerations when using this algorithm and show that it performs competitively against other lifelong learning approaches on different benchmarks. We also discuss several improvements to the algorithm and outline some future research directions for Bayesian lifelong learning.

4.1 Introduction

Deep learning provides a suite of approaches for fitting multi-layer neural networks to data. Deep neural networks have been shown to achieve remarkable prediction accuracy on several problems in important domains such as computer vision, speech processing and natural language processing. Training these models is usually done offline and separately for each individual task. Attempts to develop an effective lifelong learning process for neural networks, so that they can be trained in an incremental way to solve a set of different tasks have been hindered by either catastrophic forgetting, where models quickly forget previous tasks when trained on a new task using vanilla stochastic gradient descent (French, 1999; Goodfellow et al., 2013) or catastrophic intransigence (Chaudhry et al., 2018a), where they fail to adapt sufficiently to new data.

In this chapter, we develop a framework for performing lifelong learning in deep neural networks using probabilistic modeling and inference. Specifically, we formalize a general Bayesian framework that can handle lifelong learning in deep models in a natural and principled way. If exact inference can be performed, our framework can automatically avoid catastrophic forgetting during the learning process. The framework is also general in the sense that it can cover both discriminative and generative deep learning models, as well as task-aware (Kirkpatrick et al., 2017; Abati et al., 2020) and task-agnostic (Aljundi et al., 2019b; Zeno et al., 2021) lifelong learning settings.

As a realization of this framework, we introduce *variational continual learning* (VCL) (Nguyen et al., 2018; Swaroop et al., 2018), one of the first works to use the Bayesian framework to derive a new lifelong learning approach. VCL uses online variational inference to approximate the posterior after each task is observed. Using a previously developed method for variational inference in neural networks called Bayes by Backprop (Blundell et al., 2015), VCL effectively learns the variational parameters of the approximate posterior at each iteration by maximizing the variational lower bound. This maximization process is done using a stochastic gradient-based optimizer, such as Adam (Kingma and Ba, 2015). To further mitigate catastrophic forgetting due to successive posterior approximations, we enhance VCL with an episodic memory (called a coreset) and show how variational inference can be performed for lifelong learning in the presence of such coresets.

We also emphasize several practical considerations when using VCL, namely the importance of long training runs, the use of the local reparameterization trick (Kingma *et al.*, 2015) and sensible initialization of the variational parameters. In practice, these techniques can greatly improve model accuracy over a naive application of Bayes by Backprop. Our experiments show that VCL with these improvements performs competitively against several state-of-the-art methods on the popular Split MNIST and Permuted MNIST benchmarks for lifelong learning. We also explore the pruning effect of VCL and the role it plays in utilizing model capacity for learning a sequence of tasks.

Finally, we end the chapter by briefly discussing several improvements to VCL that have been developed in recent years, such as natural-gradient variational inference methods (Chen *et al.*, 2018; Khan *et al.*, 2018; Osawa *et al.*, 2019), generalized VCL with FiLM layers (Loo *et al.*, 2021), function-space regularization (Titsias *et al.*, 2020; Pan *et al.*, 2020) and the Stein gradient method for choosing coresets (Chen *et al.*, 2018). We also outline future research directions for using Bayesian inference to develop a realistic lifelong learning procedure for deep neural networks.

4.2 Lifelong Learning from the Bayesian Perspective

In this section, we formalize the general Bayesian approach to lifelong learning. Consider the lifelong learning setting where there is a potentially infinite stream of *data batches* $\mathcal{D}_1, \mathcal{D}_2, \mathcal{D}_3, \ldots$ that arrive over time, and we have a model with parameters $\boldsymbol{\theta}$ that we need to continually adapt to the observed data. Under the Bayesian framework, we place a prior distribution $p(\boldsymbol{\theta})$ over $\boldsymbol{\theta}$, and every time we observe a batch $\mathcal{D}_i$, we update our posterior using the likelihood $p(\mathcal{D}_i|\boldsymbol{\theta})$.

Formally, the posterior distribution $p(\boldsymbol{\theta}|\mathcal{D}_{1:T})$ after observing T data batches $\mathcal{D}_1, \mathcal{D}_2, \ldots, \mathcal{D}_T$ is obtained using Bayes' rule:

$$p(\boldsymbol{\theta}|\mathcal{D}_{1:T}) \propto p(\boldsymbol{\theta}) \prod_{i=1}^{T} p(\mathcal{D}_i|\boldsymbol{\theta}) \propto p(\boldsymbol{\theta}|\mathcal{D}_{1:T-1}) p(\mathcal{D}_T|\boldsymbol{\theta}). \qquad (4.1)$$

In the above equation, $p(\boldsymbol{\theta}|\mathcal{D}_{1:T-1}) \propto p(\boldsymbol{\theta}) \prod_{i=1}^{T-1} p(\mathcal{D}_i|\boldsymbol{\theta})$ is the previous posterior obtained after observing the first $T-1$ batches. This equation provides a natural way to handle lifelong learning using Bayesian principles. At every time step T, we only need to maintain the current posterior $p(\boldsymbol{\theta}|\mathcal{D}_{1:T})$, and we continually update this posterior in light of new observations using Bayes' rule.

The formulation for performing Bayesian lifelong learning outlined above is very general. It covers both the case where the probabilistic model is discriminative and the case where it is generative. It also covers task-aware and task-agnostic lifelong learning settings. We now outline these settings in more detail.

Discriminative model setting: For discriminative models, such as classifiers, for any $t \in \{1, 2, \ldots, T\}$, each data batch $\mathcal{D}_t$ consists of N_t labeled examples $\{(\boldsymbol{x}_t^{(n)}, y_t^{(n)})\}_{n=1}^{N_t}$, and Eq. (4.1) can be rewritten as

$$p(\boldsymbol{\theta}|\mathcal{D}_{1:T}) \propto p(\boldsymbol{\theta}|\mathcal{D}_{1:T-1}) \prod_{n=1}^{N_T} p(y_T^{(n)}|\boldsymbol{\theta}, \boldsymbol{x}_T^{(n)}), \qquad (4.2)$$

where $p(y|\boldsymbol{\theta}, \boldsymbol{x})$ is the likelihood of the parameters $\boldsymbol{\theta}$ from the labeled example $(\boldsymbol{x}, y)$.

Generative model setting: For generative models, such as deep generative models, which are often trained using the variational autoencoder approach (Kingma and Welling, 2014; Rezende *et al.*, 2014), for any $t \in \{1, 2, \ldots, T\}$, each data batch $\mathcal{D}_t$ consists of N_t unlabeled examples $\{\boldsymbol{x}_t^{(n)}\}_{n=1}^{N_t}$, and Eq. (4.1) can be rewritten as

$$p(\boldsymbol{\theta}|\mathcal{D}_{1:T}) \propto p(\boldsymbol{\theta}|\mathcal{D}_{1:T-1}) \prod_{n=1}^{N_T} p(\boldsymbol{x}_T^{(n)}|\boldsymbol{\theta}), \qquad (4.3)$$

where $p(\boldsymbol{x}|\boldsymbol{\theta})$ is the likelihood of $\boldsymbol{\theta}$ from the unlabeled example $\boldsymbol{x}$.

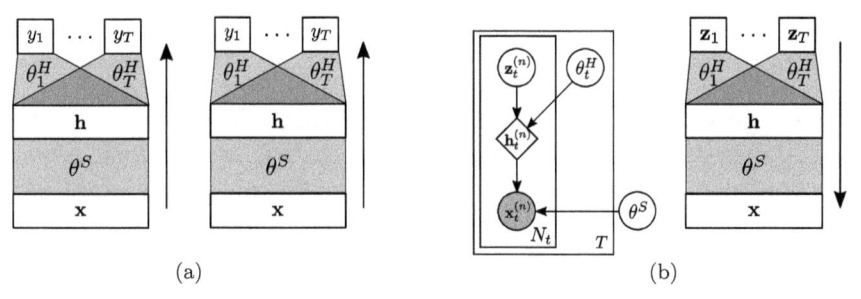

Fig. 4.1: Schematic diagrams of multi-head networks, including both the probabilistic graphical model (left) and network architecture (right), reproduced from Nguyen et al. (2018): (a) A multi-head discriminative model showing how network parameters might be shared. The lower-level network is parameterized by the variables $\boldsymbol{\theta}^S$ and is shared across multiple tasks. Each task $t \in \{1, 2, \ldots, T\}$ has its own "head network" $\boldsymbol{\theta}_t^H$ mapping onto the outputs from a common hidden representation. The full set of parameters is therefore $\boldsymbol{\theta} = \{\boldsymbol{\theta}_{1:T}^H, \boldsymbol{\theta}^S\}$. (b) A multi-head generative model (see Section 4.3.3 for details) with shared network parameters. For each task t, the head networks $\boldsymbol{\theta}_t^H$ generate intermediate-level representations from the latent variables $\mathbf{z}_t$.

Task-aware lifelong learning setting: Task-aware lifelong learning refers to the setting where the learner is aware of the time when a task switches to a new task (Kirkpatrick et al., 2017; Abati et al., 2020). This setting usually requires additional knowledge of which task each data batch $\mathcal{D}_i$ belongs to, and this knowledge will allow the learner to adjust its learning procedure accordingly. For deep neural networks, a typical way to deal with task changes in the task-aware setting is to extend the network architecture with task-specific parameters whenever a new task arrives, while maintaining a set of shared network parameters for all tasks. The task-specific parameters can simply be a classifier head (Zenke et al., 2017) or an entire network column (Rusu et al., 2016a). Figure 4.1 illustrates multi-head networks that use a separate head for each task.

The Bayesian lifelong learning formalism encapsulated in Eq. (4.1) can easily handle these types of network extensions. Specifically, whenever a new task T arrives, we add the task-specific parameters $\boldsymbol{\theta}_T^H$ to the network, resulting in a new set of parameters $\boldsymbol{\theta} = \{\boldsymbol{\theta}_{1:T}^H, \boldsymbol{\theta}^S\}$, and extend the current posterior $p(\boldsymbol{\theta}_{1:T-1}^H, \boldsymbol{\theta}^S | \mathcal{D}_{1:T-1})$ to a new posterior $p(\boldsymbol{\theta}_{1:T}^H, \boldsymbol{\theta}^S | \mathcal{D}_{1:T-1})$ with some "prior" distribution for $\boldsymbol{\theta}_T^H$. Then, we can update the posterior for the whole model, including $\boldsymbol{\theta}_T^H$, using Eq. (4.1) and data $\mathcal{D}_T$ for the new task.

Task-agnostic lifelong learning setting: In contrast to the task-aware setting above, task-agnostic or task-free lifelong learning considers a more difficult scenario where the learner does not know the exact time when a task switches (Aljundi et al., 2019b; Jerfel et al., 2019; He et al., 2020; Zeno et al., 2021; Jin et al., 2021). In other words, the learner only receives a data batch $\mathcal{D}_i$ at each time step without knowing which task $\mathcal{D}_i$ belongs to. In this case, we typically do not use multi-head architectures and maintain only a single shared network for all tasks. Our formulization in Eq. (4.1) can be applied directly to this setting without any modifications. Other Bayesian approaches, such as natural VCL (Tseran et al., 2018) or Bayesian structural adaptation (Kumar et al., 2021), can also be applied to this task-agnostic setting.

In principle, our Bayesian framework for lifelong learning presented in this section can handle continual learning naturally without suffering from catastrophic forgetting or catastrophic intransigence. Ideally, if we can always maintain the exact posterior $p(\boldsymbol{\theta}|\mathcal{D}_{1:T})$ after observing the last data batch $\mathcal{D}_T$, we would not lose any information from the prior and previous observations and thus, by design, would be able to overcome catastrophic forgetting.

In practice, however, the posterior distributions are usually intractable, especially for complex models such as neural networks. Hence, approximations are usually required to maintain the tractability of the (approximate) posteriors. Typically, after observing the data batch $\mathcal{D}_T$, an approximation method will replace the intractable true posterior $p(\boldsymbol{\theta}|\mathcal{D}_{1:T})$ by a tractable approximate distribution $q_T(\boldsymbol{\theta})$, which would be used as the prior to compute the next approximate posterior. Formally,

$$p(\boldsymbol{\theta}|\mathcal{D}_{1:T}) \approx q_T(\boldsymbol{\theta}) = \text{proj}\Big(q_{T-1}(\boldsymbol{\theta}) p(\mathcal{D}_T|\boldsymbol{\theta})\Big), \qquad (4.4)$$

where $\text{proj}(p^*(\boldsymbol{\theta}))$ is a projection operation on an intractable, unnormalized distribution $p^*(\boldsymbol{\theta})$ that returns a tractable, normalized approximate distribution. Different projection operations can be used in Eq. (4.4), such as Laplace's approximation, variational inference, moment matching or importance sampling. These projections lead to online updating methods known as Laplace propagation (Smola et al., 2004), online variational inference (Ghahramani and Attias, 2000; Sato, 2001; Broderick et al., 2013), assumed density filtering (Maybeck, 1982) and sequential Monte Carlo (Liu and Chen, 1998), respectively. In the following section, we extend the variational inference approach to online learning to support the continual learning of neural networks.

4.3 Variational Inference for Lifelong Learning

In this section, we describe VCL, a method for continual/lifelong learning with deep neural networks using variational inference. VCL can be used together with episodic memory to reduce catastrophic forgetting in these networks. We illustrate how VCL can be applied to discriminative and generative models and note some practical considerations when using the method. Finally, we show some experimental evaluations and discuss the pruning effects of this method.

4.3.1 *Variational continual learning*

VCL applies Eq. (4.4) above with a projection operation that minimizes the Kullback–Leibler (KL) divergence between the input distribution and the approximate output distribution. More specifically, the approximate posterior $q_T(\boldsymbol{\theta})$ in Eq. (4.4) is computed using

$$q_T(\boldsymbol{\theta}) = \arg\min_{q \in \mathcal{Q}} \text{KL}\left(q(\boldsymbol{\theta}) \,\|\, \frac{1}{Z_T} q_{T-1}(\boldsymbol{\theta}) p(\mathcal{D}_T|\boldsymbol{\theta})\right), \qquad (4.5)$$

where the minimization is considered over a family $\mathcal{Q}$ of distributions and Z_T is the normalizing constant of $q_{T-1}(\boldsymbol{\theta})p(\mathcal{D}_T|\boldsymbol{\theta})$, which is usually not required to solve this optimization problem. During the learning process for VCL, the zeroth approximate posterior is set to be the prior, $q_0(\boldsymbol{\theta}) = p(\boldsymbol{\theta})$, and subsequent approximate posteriors are computed using Eq. (4.5) every time a new batch of data $\mathcal{D}_T$ is received.

As discussed in Section 4.2, if we can maintain the true posterior at every iteration, we will be able to avoid catastrophic forgetting. For instance, if the family $\mathcal{Q}$ contains all of the true posteriors and assuming we can always find the true posterior when solving the minimization problem (4.5), then catastrophic forgetting can be overcome. In practice, however, we usually choose a simpler distribution family $\mathcal{Q}$, such as Gaussian distributions, leading to information loss that can cause catastrophic forgetting. To mitigate this potential forgetting problem, VCL can be extended to include a small episodic memory (Lopez-Paz and Ranzato, 2017), also called a coreset, which contains representative examples from previous batches of data to help the model refresh important information before making predictions.

This *coreset VCL* algorithm can be described as follows. At every iteration T, we observe a new batch of data $\mathcal{D}_T$ and construct a coreset C_T from $\mathcal{D}_T \cup C_{T-1}$, where C_{T-1} is the previous coreset. Instead of maintaining the approximate posterior $q_T(\boldsymbol{\theta}) \approx p(\boldsymbol{\theta}|\mathcal{D}_{1:T})$, coreset VCL maintains an approximate posterior $\tilde{q}_T(\boldsymbol{\theta}) \approx p(\boldsymbol{\theta}|\mathcal{D}_{1:T} \setminus C_T)$ that approximates the true posterior after observing all data except those in the current coreset. In the lifelong learning setting, $\tilde{q}_T(\boldsymbol{\theta})$ can be updated using the following recursion:

$$\tilde{q}_T(\boldsymbol{\theta}) = \text{proj}\big(\tilde{q}_{T-1}(\boldsymbol{\theta})\, p(\mathcal{D}_T \cup C_{T-1} \setminus C_T | \boldsymbol{\theta})\big)$$
$$= \arg\min_{q \in \mathcal{Q}} \text{KL}\left(q(\boldsymbol{\theta}) \,\|\, \frac{1}{\tilde{Z}} \tilde{q}_{T-1}(\boldsymbol{\theta})\, p(\mathcal{D}_T \cup C_{T-1} \setminus C_T | \boldsymbol{\theta})\right), \tag{4.6}$$

where $p(\mathcal{D}_T \cup C_{T-1} \setminus C_T | \boldsymbol{\theta})$ is the likelihood contribution from data points that are either not selected for the current coreset or removed from the previous coreset and $\tilde{Z}$ is the normalization constant. The recursion (4.6) ensures the approximate posterior $\tilde{q}_T(\boldsymbol{\theta})$ incorporates all non-coreset observations in the first T iterations.

When making predictions (e.g., in a classification problem) on a validation or test set, coreset VCL first refreshes the important information stored in the current coreset C_T by making the variational update

$$q_T(\boldsymbol{\theta}) = \arg\min_{q \in \mathcal{Q}} \text{KL}\left(q(\boldsymbol{\theta}) \,\|\, \frac{1}{Z}\tilde{q}_T(\boldsymbol{\theta}) p(C_T | \boldsymbol{\theta})\right), \tag{4.7}$$

and then uses $q_T(\boldsymbol{\theta})$ to make a prediction on a test input $\boldsymbol{x}^*$:

$$p(y^*|\boldsymbol{x}^*, \mathcal{D}_{1:T}) \approx \int q_T(\boldsymbol{\theta}) p(y^*|\boldsymbol{\theta}, \boldsymbol{x}^*) \mathrm{d}\boldsymbol{\theta}. \tag{4.8}$$

Using Eq. (4.7), $q_T(\boldsymbol{\theta})$ incorporates all observations $\mathcal{D}_{1:T}$ and thus can be considered an approximation of $p(\boldsymbol{\theta}|\mathcal{D}_{1:T})$. Furthermore, applying this equation immediately before making predictions enables the trained model to be less forgetful of the important information in the coreset C_T.

In principle, the coreset C_T can be constructed from $\mathcal{D}_T \cup C_{T-1}$ using various methods. The simplest methods use random sampling or K-center clustering to construct the coreset (Nguyen et al., 2018). More sophisticated methods include using Stein gradients (Chen et al., 2018), bilevel optimization (Borsos et al., 2020) or diverse sample selection (or rainbow memory) (Bang et al., 2021).

4.3.2 VCL for discriminative models

In this section, we illustrate how VCL can be used with discriminative probabilistic models. Note that the schematic diagrams for discriminative multi-head networks are shown in Fig. 4.1(a). Although we focus on vanilla VCL without the coreset here, the following formulation can be easily extended to coreset VCL.

For discriminative models, for any $t \in \{1, 2, \ldots, T\}$, each data batch $\mathcal{D}_t$ contains N_t labeled examples, $\{(\boldsymbol{x}_t^{(n)}, y_t^{(n)})\}_{n=1}^{N_t}$. The update rule for VCL in Eq. (4.5) can be rewritten as maximizing the variational lower bound to the online marginal log-likelihood:

$$q_T(\boldsymbol{\theta}) = \arg\max_{q \in \mathcal{Q}} \mathcal{L}_{\text{VCL}}^T(q(\boldsymbol{\theta})), \tag{4.9}$$

where

$$\mathcal{L}_{\text{VCL}}^T(q(\boldsymbol{\theta})) = \sum_{(\mathbf{x},y) \in \mathcal{D}_T} \mathbb{E}_{q(\boldsymbol{\theta})}\left[\log p(y|\boldsymbol{\theta}, \mathbf{x})\right] - \text{KL}(q(\boldsymbol{\theta}) \| q_{T-1}(\boldsymbol{\theta})). \tag{4.10}$$

If the family $\mathcal{Q}$ comprises Gaussian mean-field approximate posteriors, then $q(\boldsymbol{\theta}) = \prod_{d=1}^{D} \mathcal{N}(\theta_d; \mu_d, \sigma_d^2)$, where D is the number of dimensions of $\boldsymbol{\theta}$ and $\boldsymbol{\theta} = (\theta_1, \theta_2, \ldots, \theta_D)$. In this case, we can find $q(\boldsymbol{\theta})$ by maximizing Eq. (4.9) with respect to the variational parameters $\{\mu_d, \sigma_d\}_{d=1}^{D}$, which are the mean and diagonal covariance of the Gaussian. This optimization problem can be done using Bayes by Backprop (Blundell et al., 2015) and requires a Monte Carlo approximation of the expected log-likelihood term in the approximate marginal likelihood.

4.3.3 VCL for generative models

This section discusses how VCL can be applied to Bayesian generative models, specifically deep generative models, as often learned using the variational autoencoder (VAE) framework (Kingma and Welling, 2014; Rezende et al., 2014). In a standard VAE setup, we model the likelihood of each example x as $p(x|\theta) = \int p(\mathbf{x}|\mathbf{z}, \theta)p(\mathbf{z})d\mathbf{z}$, where $\mathbf{z}$ are the latent variables and $p(\mathbf{x}|\mathbf{z}, \theta)$ is defined by a neural network with parameters θ. Given an unlabeled dataset $\mathcal{D}$, the model is trained by maximizing the following variational lower bound:

$$\mathcal{L}_{\text{VAE}}(\theta, \phi) = \sum_{\mathbf{x} \in \mathcal{D}} \mathbb{E}_{q_\phi(\mathbf{z}|\mathbf{x})} \left[\log \frac{p(\mathbf{x}|\mathbf{z}, \theta)p(\mathbf{z})}{q_\phi(\mathbf{z}|\mathbf{x})} \right], \quad (4.11)$$

where $q_\phi(\mathbf{z}|\mathbf{x})$ is the approximate posterior of the latent variables (or the encoder) and ϕ denotes the variational parameters of this approximate posterior.

In the lifelong learning setting, we sequentially receive batches of data, where for any $t \in \{1, 2, \ldots, T\}$, each data batch $\mathcal{D}_t$ contains N_t unlabeled examples $\{x_t^{(n)}\}_{n=1}^{N_t}$. The schematic diagrams for multi-head generative models used in this setting are shown in Fig. 4.1(b). In VCL, we use a Bayesian VAE model where we approximate the posterior $p(\theta|\mathcal{D}_{1:T})$ with a variational distribution $q_T(\theta)$. We can then rewrite the VCL update rule (4.5) as maximizing the full variational lower bound:

$$q_T(\theta), \phi = \arg\max_{q, \phi} \mathcal{L}_{\text{VCL}}^T(q(\theta), \phi), \quad (4.12)$$

where

$$\mathcal{L}_{\text{VCL}}^T(q(\theta), \phi) = \sum_{\mathbf{x} \in \mathcal{D}_T} \mathbb{E}_{q(\theta)} \mathbb{E}_{q_\phi(\mathbf{z}|\mathbf{x})} \left[\log \frac{p(\mathbf{x}|\mathbf{z}, \theta)p(\mathbf{z})}{q_\phi(\mathbf{z}|\mathbf{x})} \right]$$

$$- \text{KL}(q(\theta)\|q_{T-1}(\theta)). \quad (4.13)$$

In this formulation, the encoder network $q_\phi(\mathbf{z}|\mathbf{x})$ is parameterized by ϕ and can be either task-specific or partly shared among different tasks. Similarly to discriminative models, we can solve the optimization problem (4.12) using backpropagation.

4.3.4 Relationship to other work

VCL can sometimes be regarded as a regularization-based approach to lifelong learning. Specifically, in the variational lower bound of Eq. (4.10) or Eq. (4.13), the first term on the right-hand side (usually called the expected log-likelihood term) favors approximate posteriors that maximize the expected log-likelihood of the training data. The second term in these equations, usually called the KL-to-prior term, serves as a regularizer that pulls the approximate posterior toward the prior (i.e., the previous approximate posterior).

Several lifelong learning methods for discriminative deep neural networks also employ a regularization-based approach that maximizes the following general objective at each iteration T:

$$\mathcal{L}^T(\boldsymbol{\theta}) = \sum_{(\mathbf{x},y)\in\mathcal{D}_T} \log p(y|\boldsymbol{\theta},\mathbf{x}) - \frac{1}{2}\lambda_T(\boldsymbol{\theta} - \boldsymbol{\theta}_{T-1})^\mathsf{T}\Sigma_{T-1}^{-1}(\boldsymbol{\theta} - \boldsymbol{\theta}_{T-1}), \tag{4.14}$$

where the matrix Σ_{T-1} controls the relative regularization strength of each element of $\boldsymbol{\theta}$ and λ_T controls the overall strength of the regularizer.

There are several instances of this regularization-based approach. For example, Laplace propagation (Smola et al., 2004) applies Laplace's approximation at each iteration, leading to the recursion $\Sigma_T^{-1} = \Phi_T + \Sigma_{T-1}^{-1}$, with $\Phi_T = -\nabla\nabla_\theta \sum_{(\mathbf{x},y)\in\mathcal{D}_T} \log p(y|\boldsymbol{\theta},\mathbf{x})|_{\boldsymbol{\theta}=\boldsymbol{\theta}_T}$ and $\lambda_T = 1$. Another popular regularization-based lifelong learning algorithm that also takes inspiration from Bayesian principles is elastic weight consolidation (EWC) (Kirkpatrick et al., 2017). This algorithm approximates $\Phi_T \approx \text{diag}\left(\sum_{(\mathbf{x},y)\in\mathcal{D}_T} (\nabla_\theta \log p(y|\boldsymbol{\theta},\mathbf{x}))^2 \big|_{\boldsymbol{\theta}=\boldsymbol{\theta}_T}\right)$ and modifies the regularization term to $\frac{1}{2}\lambda_T \sum_{t=1}^{T-1}(\boldsymbol{\theta} - \boldsymbol{\theta}_{t-1})^\mathsf{T}\Phi_t(\boldsymbol{\theta} - \boldsymbol{\theta}_{t-1})$. The regularization strengths in Σ_T^{-1} can also be computed using path integrals, as in the synaptic intelligence (SI) algorithm (Zenke et al., 2017). The performance of SI and VCL has been analyzed and shown to be correlated with the complexity of the observed tasks by Nguyen et al. (2019).

There have been several studies that take Bayesian approaches to lifelong learning. Ritter et al. (2018) used Bayesian online learning with the block-diagonal Kronecker factored approximation of the

Hessian to update their quadratic penalty for each new task. Ebrahimi et al. (2020a) trained variational Bayesian neural networks using learning rates adapted from the uncertainty induced by the probability distribution of the network's parameters. Kessler et al. (2021) employed hierarchical Indian buffet process priors for Bayesian neural networks to allow the model to use resources more effectively in lifelong learning. Wang et al. (2021) developed a method for dynamically expanding and combining model parameters to actively forget knowledge that interferes with learning new tasks. Farquhar and Gal (2018) proposed using generative adversarial networks to generate data for old tasks that can be used together with VCL. Zeno et al. (2021) considered the task-agnostic lifelong learning setting and derived fixed-point updates for variational Bayesian neural networks. Chen et al. (2021) developed a generative regularization approach to prevent catastrophic forgetting, and finally, Bayesian lifelong learning with non-stationary data was considered by Kurle et al. (2020).

4.3.5 *Practical considerations for VCL*

We now discuss applications of VCL to discriminative lifelong learning benchmarks. As discussed in Section 4.3.2, we use Bayes by Backprop (Blundell et al., 2015) to optimize the means and variances $\{\mu_d, \sigma_d\}_{d=1}^D$ of our mean-field Gaussian approximating family, optimizing the objective function (4.10). This involves Monte Carlo sampling of the expected log-likelihood term, and we use Adam (Kingma and Ba, 2015) to optimize the function.

There are several details that, although seemingly technically trivial, are practically important and will greatly improve results over naively applying Bayes by Backprop for VCL. Crucially, we optimize for a much longer time than previous works and do not early-stop. Although progress can sometimes appear to stall during optimization, in reality, progress is just extremely slow. We can also speed up the algorithm's convergence rate using two simple techniques: using the local reparameterization trick and improving the initialization of the variational parameters.

Local reparameterization trick: We can employ the local reparameterization trick (Kingma et al., 2015) during Monte Carlo sampling of the likelihood term. Specifically, instead of sampling each

(Gaussian) weight independently, we sample the pre-activation latent variables just before each neuron's nonlinearity (this latent variable is a linear combination of the neuron's input weights). This leads to two improvements: (i) It reduces the variance of stochastic gradients during sampling, and (ii) it is marginally quicker as we sample fewer random variables. The first improvement is particularly important as it speeds up convergence drastically, substantially reducing the number of epochs required for convergence.

Initialization: Another important trick is to improve the initialization of the weights of our neural network when running the optimizer. We experimented with (i) initializing the mean-field Gaussian weights by setting the means at the maximum likelihood solution of a deterministic neural network and setting the variances to be small, (ii) initializing the means to be small and random and setting the variances to be small, and (iii) initializing at the means and variances of the prior. We find that using (ii) improves the convergence speed and reduces the standard deviation in final accuracy across many runs. Intuitively, this is because initializing randomly allows the network to quickly learn the best trade-off between the new task's data (the expected log-likelihood term) and information from previous tasks (the KL-to-prior term). Specifically, the means are randomly initialized to be of the order 10^{-1} and the variances are initialized to be of the order 10^{-3}.

4.3.6 *Experimental evaluations of VCL*

In this section, we show some experimental evaluations of VCL after incorporating the practical tips in Section 4.3.5. We consider the following two lifelong learning benchmarks, both of which use the popular MNIST handwritten digit recognition dataset. The code to run all experiments is available at https://github.com/nvcuong/variational-continual-learning.

Split MNIST: In this benchmark, we have to sequentially solve five binary classification tasks from the MNIST dataset: {0v1}, {2v3}, {4v5}, {6v7} and {8v9}. The challenge in Split MNIST is to obtain good performance on new tasks while retaining performance on old ones. We assume a *task-aware* setting and use the previously discussed multi-head setup. We train a one-hidden-layer

Table 4.1: Final average test accuracy on Split MNIST for various methods. Methods with an asterisk (*) use some sort of episodic memory. Results marked with a double asterisk (**) were taken or read from a graph in the respective papers.

Method	Hidden layer size	Final average test accuracy
VCL	{200}	98.5 ± 0.4%
*VCL + 40 random coreset	{200}	98.2 ± 0.4%
EWC (Kirkpatrick et al., 2017)	{200}	63.1%
Laplace Propagation (Smola et al., 2004)	{200}	61.2%
SI (Zenke et al., 2017)	{200}	98.9%
Vadam VCL (Tseran et al., 2018)	{256, 256}	99.2%**
UCL (Ahn et al., 2019)	{256, 256}	99.7%**

neural network with 200 units and ReLU activation functions and use a standard normal prior for the first task. We train for 600 epochs (with a batch size of 256) using an Adam learning rate of 5×10^{-3}, and we report the mean and standard deviation over 10 runs. The results are summarized in Table 4.1. Without coresets, we achieve a final test accuracy of 98.5 ± 0.4%, and with a coreset of 40 randomly chosen data points per task, we achieve 98.2 ± 0.4% (coresets are not required to improve results on this benchmark). These are improvements compared to previous lifelong learning methods, such as EWC (Kirkpatrick et al., 2017), which achieves 63.1%, and Laplace propagation (Smola et al., 2004), which achieves 61.2%.

Permuted MNIST: This benchmark consists of tasks received sequentially, each of which is the standard (10-way) MNIST classification task, with the pixels having undergone a fixed permutation randomly selected for each task. Ideally, a network with two or more hidden layers would use the lower layer(s) to "de-permute" the images and the higher layer(s) to solve MNIST, which is then constant between tasks (we use a single-head setup as this is a task-agnostic setting). We train a two-hidden-layer model with 100 units in each layer and ReLU activation functions, again using a standard normal prior for the first task. We train for 800 epochs (with a batch size of 1024), using an Adam learning rate of 5×10^{-3}, and we report the mean and standard deviation over five runs. The results are summarized in Table 4.2. Without coresets, VCL achieves a final

Table 4.2: Final average test accuracy on Permuted MNIST for various methods (results taken from respective papers). A hidden layer size of $\{n_1, n_2\}$ indicates two hidden layers, the lower hidden layer having n_1 hidden units and the upper hidden layer having n_2 units (followed by a softmax over the 10 MNIST classes). Methods with an asterisk (*) use some sort of episodic memory. Results marked with a double asterisk (**) were read from a graph in the source paper. Results with a dagger ($\dagger$) are from Chaudhry et al. (2018b).

Method	Hidden layer size	Number of tasks	Final average test accuracy
VCL	{100, 100}	10	$93 \pm 1\%$
*VCL + 200 random coreset	{100, 100}	10	$94.6 \pm 0.3\%$
Kronecker-factored Laplace	{100, 100}	50	90%
(Ritter et al., 2018)	{100, 100}	10	96%**
EWC (Kirkpatrick et al., 2017)	{2000, 2000}	10	97%
	{100, 100}	10	84%
SI (Zenke et al., 2017)	{2000, 2000}	10	97%
	{100, 100}	10	86%
CLNP (Golkar et al., 2019)	{2000, 2000}	10	$98.42 \pm 0.04\%$
	{100, 100}	10	95.8%
*A-GEM (Chaudhry et al., 2018b)	{256, 256}	20	$89.1 \pm 0.14\%$
	{256, 256}	10	92.3%**
BLLL-REG (Ebrahimi et al., 2020a)	{100, 100}	10	92.2%
*GEM	{256, 256}	20	$89.5 \pm 0.48\%^\dagger$
(Lopez-Paz and Ranzato, 2017)	{256, 256}	10	93.1%**†
	{100, 100}	20	80%
Riemannian Walk	{256, 256}	20	$85.7 \pm 0.56\%^\dagger$
(Chaudhry et al., 2018a)	{256, 256}	10	91.6%**†
Progressive NNs	{256, 256}	20	$93.5 \pm 0.07\%^\dagger$
(Rusu et al., 2016a)	{256, 256}	10	94.6%**†

average test accuracy of 93 ± 1%, and with coresets, VCL achieves 94.6 ± 0.3%. For reference, training the same network but seeing all the data together (batch mode) has an accuracy of 97% (this is an upper bound on the performance possible with this model and inference scheme).

4.3.7 *Discussion*

We now look into how VCL continually learns tasks, exploring how it uses its model capacity, primarily by looking at the learned weights in the model.

Split MNIST: We first consider the model trained without coresets (although exactly the same effects happen with coresets). When we look at weight values into and out of each unit after training, we find that only one unit is being used in each of the five tasks, despite having 200 units in the single layer; the remaining units are pruned out as part of the optimization process. The five active (unpruned) units are plotted in Fig. 4.2. This effect is similar to that observed by Trippe and Turner (2017), with entire units pruned out, as opposed to just individual weights. The pruned units appear to have input weights at or near the prior (standard normal Gaussian) and output weights near a delta function (zero mean, small variance), therefore minimizing their effect on the output prediction. Removing all pruned units from the network does not change the network's accuracy.

This pruning effect seems to be due to the choice of the inference scheme. Intuitively, the pruning effect can be explained by looking at the optimization function (4.10). By reducing the effect of a unit on the output prediction (e.g., setting output weights to have a zero mean and small variance), the input weights to the unit can be set to their prior. The increase in the KL-to-prior term due to the small variance of the output weights is offset by the reduction in the KL-to-prior term from the numerically higher input weights. Provided the expected log-likelihood term does not change too much, a pruned solution is therefore more optimal. We now focus on what our pruned solutions reveal about how our model approaches lifelong learning tasks and debate whether the pruning effect is a feature or a bug for lifelong learning.

Lifelong Learning for Deep Neural Networks with Bayesian Principles 67

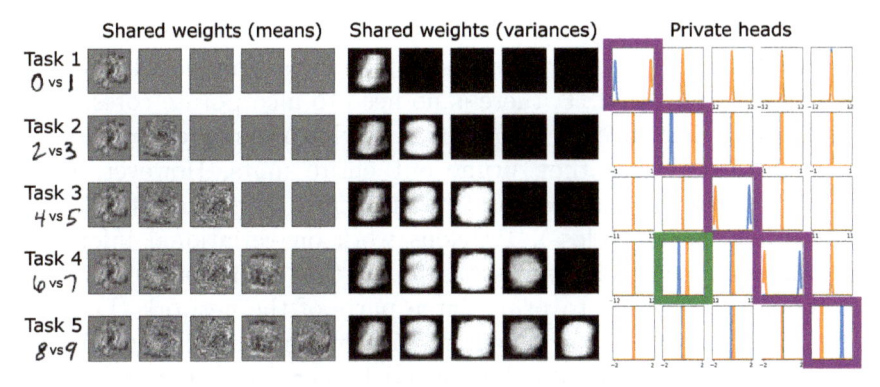

Fig. 4.2: The active units learned for Split MNIST without coresets. Rows correspond to stages of lifelong learning: (left) Means, (center) Variances, (right) Output weights for each task's two classes. Exactly the same effect is observed when incorporating coresets.

As the model only uses one hidden unit per task, it has learned to use a fraction of its total capacity to successfully learn binary classifiers in the Split MNIST experiment. The high accuracies indicate that this is an efficient use of model capacity. The remaining unused units can be used for other tasks that we may see in the future. This implies that pruning is a beneficial feature for lifelong learning, forcing the model to efficiently use its capacity.

Additionally, the pruning effect allows us to see some forward and backward transfers (see Fig. 4.2), which are both important qualities in a good lifelong learning solution. Forward transfer is visible when previous tasks' active units have non-zero weights for subsequent tasks. For example, unit 2, which was learned after task 2 (classifying digits {2v3}), has non-zero output weights after task 4 (classifying {6v7}). The model therefore uses some information about the task {2v3} in solving the task {6v7}. This is highlighted in green on the right-hand side of Fig. 4.2. Although less visible in the plots, there is also backward transfer in the same units: unit 2's input weights change slightly after training on task 4 ({6v7}), potentially changing accuracy on task 2. In this case, however, any backward transfer does not result in a different accuracy; this could be because there is no potential for improvement given the high accuracies involved.

The exact same effects are seen when we train the model with coresets: The same pruning effect occurs (Fig. 4.2 is similar when trained with coresets), and similar accuracies are obtained (within one standard deviation). This shows that for a simple task such as multi-head Split MNIST, there is no need to incorporate coresets.

This pruning effect is also similar to that considered by Golkar et al. (2019), where they prune out entire units. However, they have hyperparameters that control the degree of pruning (and corresponding accuracy loss). Previous work on variational inference for Bayesian neural networks has found that variational inference methods can be used to prune large parts of the network (Louizos et al., 2017), and we show here how this pruning is done over units as opposed to weights (see also Trippe and Turner, 2017). In comparison to that adopted by Golkar et al. (2019), our method automatically prunes out entire units and is able to reuse the units for both forward and backward transfers.

Permuted MNIST: We now look into how the two-hidden-layer model learns the Permuted MNIST task (results in Table 4.2). We first consider the model trained without coresets. We find that there is still pruning within the network. The numbers of active (unpruned) units after training on each task are summarized in Fig. 4.3.

There are more active units in Permuted MNIST than there were in Split MNIST, perhaps due to the more difficult nature of Permuted MNIST (classifying between 10 digits, as opposed to between 2).

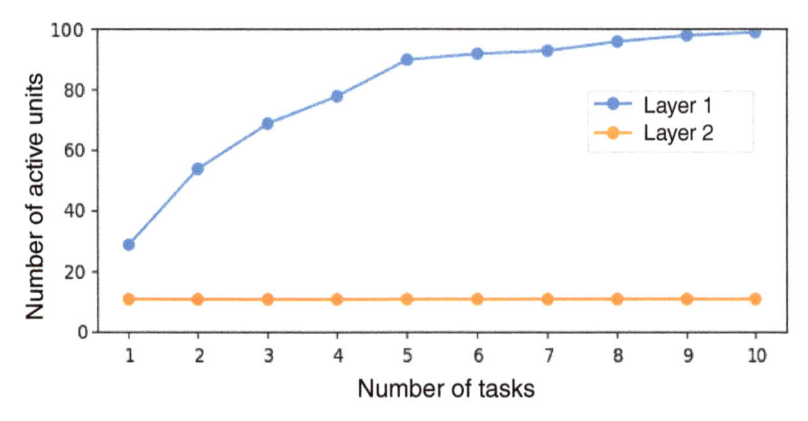

Fig. 4.3: Numbers of active units per hidden layer after each task in Permuted MNIST without coresets. Exactly the same effect is observed when incorporating coresets.

However, only 11 units are used in the second hidden layer, with the remaining 89 units pruned out. Additionally, the output weights on these 11 units do not change between tasks. This confirms that the hidden layers effectively de-permute the images, allowing the output weights to just classify between the 10 digits.

Beyond reusing the upper-level weights, there is not much evidence of forward or backward transfer. We should expect this from Permuted MNIST because the network trains on all MNIST digits on the first task itself, hence already learning the "best" way to classify between MNIST digits. Any subsequent permuted images have little room to improve upon this. Instead, the remaining focus of Permuted MNIST seems to be on ensuring we use available model capacity as efficiently as possible. Increasing the model capacity improves results. Training a network with 250 units in the lower hidden layer (instead of 100) improves the final average test accuracy to 95.5% (10 tasks).

Incorporating coresets also improves results. However, the number of active units (plotted in Fig. 4.3) remains the same. Instead, training VCL with coresets (which can be viewed as changing the order in which the model trains on data or changing the schedule with which we visit training data) appears to reinforce previous tasks' images: It lowers forgetting in the network.

Table 4.2 summarizes some recent works' results on Permuted MNIST. As can be seen, different papers use different numbers of hidden units in their hidden layers and test on different numbers of tasks. However, as the findings in this chapter indicate, Permuted MNIST primarily tests for model capacity, highlighting how we cannot faithfully compare results when model capacity and number of tasks differ. We propose that when using Permuted MNIST as a benchmark in lifelong learning, model capacity is kept fairly limited (for example, two hidden layers with either 100 or 256 hidden units each), and the number of tasks is kept high (minimum 10, possibly more). Of all methods, those proposed by Ritter *et al.* (2018) and Golkar *et al.* (2019) achieve significantly superior results, as they use a relatively small network and test over many tasks. Note that 2,000 units in two hidden layers is an extremely large model for 10 tasks: As Golkar *et al.* (2019) note, they can achieve high (as good as single-task) performance having pruned out a large proportion of their network. With such a large model capacity, this benchmark no longer tests any desiderata in lifelong learning aside from avoiding catastrophic forgetting, which many other benchmarks can also do.

Many studies often use Permuted MNIST to demonstrate how their method (and other baseline methods) exhibit some of lifelong learning's desiderata. For example, they show some form of resistance to forgetting or efficient use of model capacity via metrics or plots. In this case, comparing the final average test accuracy is no longer so important; however, we believe far better comparisons would be achieved if a more challenging hidden layer size and number of tasks were used.

4.4 Improvements to VCL

There has been follow-up work that uses the core ideas that we have discussed above and looks at generalizing and improving them. We briefly discuss them in this section.

Natural-gradient variational inference: One strand of work optimizes the variational objective function (4.5) with natural gradients instead of using Bayes by Backprop to optimize for the variational parameters (Chen *et al.*, 2018; Khan *et al.*, 2018; Osawa *et al.*, 2019). Natural-gradient update steps are a principled way of incorporating the information geometry of the distribution being optimized (Amari, 1998). By incorporating the geometry of the distribution, we expect to take gradient steps in much better directions, speeding up gradient-based optimization. By using such natural-gradient variational inference, the objective (4.5) is therefore optimized quicker with the same performance on MNIST benchmarks (Osawa *et al.*, 2019; Eschenhagen, 2019). The faster optimization has also allowed the VCL objective function to be scaled to larger benchmarks (Eschenhagen, 2019) and architectures.

Generalized VCL and FiLM layers: It is possible to temper the KL-to-prior term in the variational objective function (4.10). Loo *et al.* (2021) showed that this modification to VCL (called *generalized VCL*) recovers online EWC (Schwarz *et al.*, 2018) as a limiting case, allowing for interpolation between the two approaches. This leads to theoretical generalization and increased performance on several benchmarks.

Loo et al. (2021) also introduced task-specific FiLM layers (Perez et al., 2018) to take advantage of and reduce pruning in variational Bayesian neural networks, finding that this also leads to improved performance. These FiLM layers linearly modulate features in a neural network, and making these additional parameters task-specific allows useful features for a task to be amplified and inappropriate ones to be ignored. This improves the pruning effect by allowing FiLM layers to prune units instead of relying on the output weights of a unit to become zero-mean low-variance distributions. Loo et al. (2021) found that FiLM layers reduce pruning and improves the performance of VCL (and generalized VCL) on many benchmarks.

Function-space regularization: Over a sequence of many tasks, independent regularization of weights can still lead to forgetting. Instead, we are ultimately interested in neural network outputs or predictions, and we want to maintain these predictions over the course of many tasks. This has led to the idea of regularizing the function space of neural networks directly, instead of only regularizing in the weight space (Titsias et al., 2020; Pan et al., 2020). These methods store input–output pairs (the memorable examples), similar to coresets in VCL, but regularize them directly in the loss function. They are derived from the variational objective function and use the same core idea of variational updates for continual learning. They show significantly better performance than VCL on larger-scale problems and are a promising direction for future research.

Improving coreset selection with Stein gradients: In the original version of VCL, coresets are chosen without knowledge of the approximate posterior using, for example, random sampling or K-center clustering. Chen et al. (2018) proposed an improvement to VCL that uses Stein gradients to construct the coresets. This method iteratively updates a coreset and moves it closer to samples from the posterior distribution, thus allowing the coreset to be constructed using the approximate posteriors without changing the inference procedure. Coresets constructed using Stein gradients are shown to outperform random and K-center coresets on the Permuted MNIST benchmark.

4.5 Conclusions and Future Directions

This chapter introduced the Bayesian approach to lifelong learning for deep neural networks. This Bayesian framework provides a principled way to tackle lifelong learning and has great potential for future research. One important question is how to scale Bayesian inference, in general, and lifelong learning, in particular, to a large dataset, such as ImageNet (Deng et al., 2009), and for very deep neural networks. A good solution to this question would enable the applications of Bayesian methods to several state-of-the-art deep models in computer vision or natural language processing that contain several million parameters. Another important direction for future explorations is to improve function-space regularization approaches, potentially by using better posterior approximations and better memorable example selection mechanisms, and scale these approaches to larger datasets.

© 2024 World Scientific Publishing Company
https://doi.org/10.1142/9789811286711_0005

Chapter 5

Generative Replay-Based Continual Zero-Shot Learning

Chandan Gautam[*,§], Sethupathy Parameswaran[†,∥], Ashish Mishra[‡,¶] and Suresh Sundaram[†,**]

[*]*Institute for Infocomm Research, Agency for Science, Technology and Research (A*STAR), Singapore*
[†]*Indian Institute of Science, Bangalore, India*
[‡]*Indian Institute of Technology Madras, Chennai, India*
[§]*gautam_chandan@i2r.a-star.edu.sg*
[∥]*sethupathyp@iisc.ac.in*
[¶]*mishra@cse.iitm.ac.in*
[**]*vssuresh@iisc.ac.in*

Abstract

Zero-shot learning (ZSL) is a new paradigm to classify objects from classes that are not available at training time. ZSL methods have attracted considerable attention in recent years because of their ability to classify unseen/novel class examples. Most of the existing approaches on ZSL work when all the samples from seen classes are available to train the model, which does not suit real life. In this chapter, we tackle this hindrance by developing a generative replay-based continual ZSL (GRCZSL). The proposed method endows traditional ZSL with the ability to learn from streaming data and acquire new knowledge without forgetting the previous tasks' gained experience. We handle catastrophic forgetting in GRCZSL by replaying the synthetic samples of seen classes which have appeared in the earlier tasks. These synthetic samples are synthesized using the trained conditional variational autoencoder (VAE) over the immediate past task. Moreover, we only require the current and immediately previous VAE at any time for training and testing. The proposed GRZSL method is developed for a single-head setting of continual learning, simulating a real-world problem setting. In this setting, task identity is given during training but unavailable during testing. The GRCZSL performance is evaluated on five benchmark datasets for the generalized setup of ZSL with fixed and dynamic (incremental class) settings of continual learning. The existing class setting presented recently in the literature is not suitable for a class-incremental setting. Therefore, this study proposes a new setting to address this issue. Experimental results show that the proposed method significantly outperforms the baseline and the state-of-the-art method and is more suitable for real-world applications.

5.1 Introduction

Image classification using deep learning has exhibited satisfactory performance for the fully supervised learning task. However, these conventional deep learning algorithms rely heavily on a large number of labeled visual samples. Moreover, it is challenging to collect labeled samples for every class in the real world. This leads to the problem of recognizing samples from unseen/novel classes when there are no visual samples present in the training data. It is an easier task to recognize unseen classes based on their descriptions by humans, who can do this based on their imagination capability, which they have learned from their past experiences. The solution approach in the machine learning literature is popularly known as zero-shot learning (ZSL) (Farhadi *et al.*, 2009; Lampert *et al.*, 2013; Xian *et al.*, 2017). ZSL recognizes unseen classes based on their semantic associations with seen classes. The semantic information could be human-annotated attributes (Song *et al.*, 2018; Morgado and Vasconcelos, 2017; Annadani and Biswas, 2018) or text descriptions (Lei Ba *et al.*, 2015; Elhoseiny *et al.*, 2017), such as the main color of the class or word vectors (Frome *et al.*, 2013a; Zhang *et al.*, 2017; Changpinyo *et al.*, 2017) based on class names. ZSL has recently received a surge of interest in the research community, and various methods have been developed to tackle this problem. Despite the development of various ZSL methods, it is difficult to continuously learn from the sequence of tasks without forgetting the previously accumulated knowledge. Leveraging past experience and acquiring new knowledge from streaming data is known as continual/lifelong learning. This learning strategy merged with ZSL enables us to develop a continual learning framework, i.e., continual zero-shot learning (CZSL).

In conventional ZSL, the purpose of the trained model is to identify the unseen classes in the following two kinds of settings during testing: (i) disjoint setting, in which the classification search space consists of only unseen classes, and (ii) generalized setting, in which the classification search space consists of both seen and unseen classes. Here, seen and unseen classes are completely disjoint, i.e., seen classes ∩ unseen classes = ϕ. A generalized setting where a model doesn't know during testing whether an upcoming image belongs to a seen or unseen class is more realistic than the disjoint setting. This chapter also follows the generalized setting as it is closer to the real-world problem. Generally, ZSL can be performed in two ways: (i) by learning to map from the visual space to the semantic space and vice versa using a mapping function, or (ii) by generating synthetic samples for unseen classes using a generative model. However, these ZSL methods are not effective in handling the CZSL problem.

Most recently, only a handful of researchers have made an effort to tackle this CZSL problem (Wei *et al.*, 2020; Skorokhodov and Elhoseiny, 2021). CZSL has been developed for two types of settings, i.e., multi-head (Wei *et al.*, 2020) and single-head (Skorokhodov and Elhoseiny, 2021) settings. Task identity is known throughout training and testing in a multi-head setting, and each task has its own classifier. In a single-head setting, task identification is unknown at inference time, and a shared classifier is used among all tasks. It can be argued that the assumption of a multi-head setting is not practical and not a realistic setting, while the single-head setting is closer to the real-world scenario. Thus, we develop a single-head setting-based continual learning method for generalized ZSL. The existing single-head setting (Skorokhodov and Elhoseiny, 2021) is not suitable for class-incremental learning as it assumes that the semantic information of all seen and unseen classes is known at the very first task. This is an unrealistic assumption. Therefore, we propose a new setting to address this issue. This proposed CZSL setting also helps in analyzing the model's performance compared to the offline mode (i.e., upper bound). In this chapter, to develop a CZSL method, we need to deal with two aspects: continual learning and ZSL. Continual learning is dealt with using a generative replay strategy, and ZSL is dealt with using a generative model, the conditional variational autoencoder (CVAE) (Mishra *et al.*, 2018).

The overall contribution of this work can be summarized in the following points:

(1) The existing generative replay-based method can help perform a traditional classification task (i.e., classification of seen classes) in a continual learning manner. However, this chapter develops a generative replay-based continual ZSL (GRCZSL), which can also identify the unseen class by only using the class's semantic information.
(2) GRCZSL uses the immediate previous decoder model to handle catastrophic forgetting in CZSL, which makes it quite efficient. It does not require storing anything other than the previous decoder model.
(3) The proposed method has been experimented on two types of CZSL settings: (i) a fixed setting, where all the classes of the tasks presented to the model so far are taken as seen and all the classes of the remaining tasks are taken as unseen; and (ii) a dynamic setting, where each task contains new seen and unseen classes that were not contained in the previous tasks presented to the model. The first setting (fixed) has been recently developed in the literature (Skorokhodov and Elhoseiny, 2021). The second setting (dynamic) is proposed for two purposes, addressing the shortcomings of the fixed setting: (a) to assess the class-incremental ability of the CZSL method as it is not possible to evaluate the class-incremental strategy on the fixed setting (Skorokhodov and Elhoseiny, 2021); (b) to analyze the performance of the CZSL model compared to an offline model as the fixed setting does not have any unseen class after the last task; however, in a dynamic setting, the train–test split of the last task and the split of the standard ZSL are identical.
(4) As GRCZSL does not require task information at the inference time, it performs task-agnostic prediction and is suitable for the single-head setting. It shares the same model among all tasks.
(5) We conduct experiments on the five standard ZSL datasets by splitting them among different tasks for CZSL. Moreover, two kinds of CZSL settings, namely fixed and dynamic settings, are used for the experiment. GRCZSL outperforms the existing state-of-the-art method (Skorokhodov and Elhoseiny, 2021)

by more than 2% and 4% for the CUB and SUN datasets in the fixed setting, respectively.

The remaining chapter is organized as follows: Section 5.2 contains a literature survey, and Section 5.3 presents the proposed method. Experimental results and an ablation study are discussed in Section 5.4, and the last section (i.e., Section 5.5) concludes this chapter.

5.2 Related Work

This section contains related work in three parts: (i) ZSL, (ii) continual learning and (iii) CZSL.

5.2.1 *Zero-shot learning*

ZSL was initially introduced by Lampert *et al.* (2009) for attribute-based classification and is considered a disjoint setting of ZSL. One common approach to solving the ZSL problem can be categorized as prototypical, where a prototype vector for each class is computed. The class prototype vector can be computed in the following three ways: (i) Map the class attribute vector to the visual space and use these mapped attributes as class prototype vectors (Li *et al.*, 2019b); (ii) map the visual features to the attribute space and use the class attribute vectors as the prototype (Cacheux *et al.*, 2019); or (iii) map both the visual features and class attribute vectors to a common space, and use the mapped attributes as the class prototype vector (Cacheux *et al.*, 2019). The mapping function is learned using the seen classes data. Once the mapping function is learned, it is used to compute the class prototype vectors for the unseen classes as well. After obtaining the class prototype vectors, each visual feature is assigned the class of the class prototype vector to which it is most similar based on a similarity measure, such as cosine similarity. Another common approach to solving the ZSL problem is to learn a compatibility function. ALE (Akata *et al.*, 2013), DEVISE (Frome *et al.*, 2013b), SJE (Akata *et al.*, 2015) and EZSL (Romera-Paredes and Torr, 2015) learn a bilinear form as a compatibility function. Kodirov *et al.* (2017) developed a method that involves mapping visual to semantic and again mapping from mapped semantic space to visual space using an autoencoder loss to improve the reconstruction ability of the model. LATEM (Xian et al., 2016) and CMT (Socher et al., 2013) encode an additional nonlinearity in the compatibility framework.

In recent years, ZSL has employed generative models, such as a variational autoencoder (VAE) (Kingma and Welling, 2013) and a generative adversarial network (Goodfellow et al., 2014; Arjovsky et al., 2017), to learn the samples from attributes. These methods have received significant attention in the literature due to their ability to generate high-quality synthetic samples. In the literature, various generative method-based ZSL and GZSL methods have been proposed in recent years (Kumar Verma et al., 2018; Verma et al., 2021b; Xian et al., 2019; Huang et al., 2019; Sariyildiz and Cinbis, 2019; Mishra et al., 2018; Felix et al., 2018; Schonfeld et al., 2019; Chou et al., 2021; Keshari et al., 2020), and they have significantly outperformed non-generative ZSL methods.

First, these generative methods generate synthetic samples for seen and unseen classes, and then a supervised classifier is trained based on these synthetic samples. In this approach, bias toward seen classes is handled effectively, as the required number of samples for seen and unseen classes can be generated using a trained generative model. In this chapter, we use a generative method, i.e., VAE, as the base method.

5.2.2 Continual learning

There are two main issues in continual learning which need to be addressed during learning: catastrophic forgetting and intransigence (Chaudhry et al., 2018a). The literature of continual learning can be broadly divided into three categories: (i) regularization-based methods (Kirkpatrick et al., 2017; Rebuffi et al., 2017b; Chaudhry et al., 2018a), (ii) parameter isolation methods (Rusu et al., 2016a; Xu and Zhu, 2018b; Mallya and Lazebnik, 2018; Mallya et al., 2018; Aljundi et al., 2017) and (iii) replay-based methods (Lopez-Paz and Ranzato, 2017; Shin et al., 2017; Chaudhry et al., 2018b; Hayes et al., 2018a; Chaudhry et al., 2019). A few regularization-based methods handle catastrophic forgetting by storing previously learned networks and performing knowledge distillation using them (Rebuffi et al., 2017b).

Conversely, other regularization-based methods control the parameters and discourage the modification of the learned parameters on previous tasks to avoid catastrophic forgetting (Kirkpatrick et al., 2017). Parameter isolation methods assign different model parameters to each task to avoid catastrophic forgetting. When there are no restrictions on architecture size, one can assign new branches for new tasks while fixing the parameters of the previous tasks (Rusu et al., 2016a; Xu and Zhu, 2018b). As another option, the architecture is kept static, and a part of the architecture is assigned to each task (Mallya and Lazebnik, 2018; Serra et al., 2018). On the other hand, replay-based methods either replay the samples from the past task while training the current task or use those samples to provide the constraint on the optimization problem in such a way that loss on the previous tasks does not increase (Lopez-Paz and Ranzato, 2017; Chaudhry et al., 2018b). Since replay-based methods generally outperform regularization-based methods in the literature (Shin et al., 2017), we have also employed a generative replay-based strategy for handling catastrophic forgetting for CZSL.

5.2.3 Continual zero-shot learning

ZSL from streaming data is an unexplored area of research. Only a handful of studies are available for CZSL (Wei et al., 2020; Skorokhodov and Elhoseiny, 2021). First, Chaudhry et al. (2018b) discussed the possibility of continual learning using average gradient episodic memory (A-GEM) in a multi-head setting. Recently, Wei et al. (2020) developed a CZSL method using a generative model and knowledge distillation. However, this CZSL method is only compatible with multi-head settings. For the sake of comparison, we illustrate the differences between the study by Wei et al. (2020) and our work in Table 5.1. Most recently, A-GEM has been further developed for a single-head setting (Skorokhodov and Elhoseiny, 2021) using episodic gradient memory.

5.3 Proposed Method: Generative Replay-Based Continual Zero-Shot Learning

This section presents the problem formulation for generalized CZSL and then describes the CVAE architecture used with the proposed

Table 5.1: Comparison between our work and the study by Wei *et al.* (2020).

Lifelong ZSL (Wei *et al.*, 2020)	Our work
1. The mentioned CZSL setting is only feasible for a multi-head setting, i.e., a separate classifier is used for each task. Hence, task boundary must be available at the testing time, which is not practical in realistic cases.	We used a single-head CZSL setting and also proposed another kind of single-head CZSL setting which is feasible for class-incremental learning, as an existing single-head CZSL setting assumes all classes information is known *a priori*. Here, task-boundary information is not required during testing. Further, the proposed setting is more general and can be used to evaluate other CZSL methods.
2. The method used to solve the CZSL problem has used different encoders for each task's attribute and different classifiers for each task.	As our method is based on a single-head setting, only one VAE and classifier is used, which are shared among all tasks.
3. This study does not use any kind of replay strategy. The authors freeze the intermediate layer and use knowledge distillation during training to minimize catastrophic forgetting.	Our proposed method has used replay strategy with knowledge distillation and a basic VAE model for ZSL. Further, our proposed framework for CZSL will be suitable for any kind of generative model-based ZSL.
4. This method and setting cannot be deployed for a class-incremental setting.	The proposed method can handle a class-incremental learning scenario.

GRCZSL method in the subsequent section. In the final section, incremental-class learning of the GRCZSL method is explained.

5.3.1 Problem Formulation

Let us assume that the number of training samples for the tth task is denoted as n_{tr}, and the CZSL is trained on a data stream $\mathcal{D}_{\tau r}^t = \{(x_i^t, \iota_i^t, y_i^t, a_i^t)_{i=1}^{n_{\text{tr}}}\}$. Here, x_i^t denotes feature vector, ι_i^t denotes task identity, y_i^t denotes class label, and a_i^t denotes class attribute information for the ith sample of the tth task. This class attribute information is a numerical vector obtained either by human annotations or a word vector. This class attribute information is essential for implementing CZSL. Further, CZSL is tested on the data stream $\mathcal{D}_{TS}^t = \{(x_i^t, y_i^t)_{i=1}^{n_{ts}}\}$. Here, the feature vector and class label are represented as x_i^t and y_i^t for the ith sample of the tth task, respectively. The number of test samples for the tth task is represented as n_{ts}. Here, the objective is to develop an algorithm which can perform CZSL.

5.3.2 Architecture of the conditional variational autoencoder

For developing the proposed method of GRCZSL, we employed a VAE (Kingma and Welling, 2013). VAE is a type of generative model which approximates the distribution of the latent space z to that of the input data x, i.e., assume that there exists a hidden variable, say z, which can generate an observation x. Since one can only observe x, we are interested in the prior $p(z|x)$ given by

$$p(z|x) = \frac{p(x|z)p(z)}{p(x)}. \tag{5.1}$$

However, estimating $p(x)$ is quite difficult, and it is often an intractable distribution. Hence, using variational inference, we approximate $p(z|x)$ using the parametrized distribution $q_\Phi(z|x)$. Since we want $q_\Phi(z|x)$ to be similar to $p(z)$, we can estimate the parameters of $q_\Phi(z|x)$ by minimizing the Kullback–Leibler (KL) divergence (Kullback and Leibler, 1951) loss ($\mathcal{L}_{KL}$) between the two distributions, which is computed as follows:

$$\mathcal{L}_{KL} = -\mathrm{KL}(q_\Phi(z|x) \| p_\theta(z)). \tag{5.2}$$

Along with this KL divergence loss, VAE also minimizes the reconstruction loss ($\mathcal{L}_{Re}$):

$$\mathcal{L}_{Re} = \mathbb{E}_{q_\Phi(z)}\left[\log p_\theta(x|z)\right]. \tag{5.3}$$

Overall, VAE consists of the following loss, which is also called the variational lower bound for VAE:

$$\mathcal{L}_{VAE} = -\mathrm{KL}(q_\Phi(z|x) \| p_\theta(z)) + \mathbb{E}_{q_\Phi(z)}\left[\log p_\theta(x|z)\right]. \tag{5.4}$$

Here, $p_\theta(x|z)$ can be modeled as a decoder with parameters θ mapping latent space onto data space, and $q_\Phi(z|x)$ can be modeled as an encoder mapping data space onto latent space.

The encoder gives the probability distribution $q(z|x, a)$, which is assumed to be an isotropic Gaussian distribution. We model the encoder using a neural network. The encoder takes the concatenated input feature x and the respective attribute feature a as input and gives the parameter vector of the Gaussian (μ_x, Σ_x) as output. The decoder takes the concatenated z and a and tries to reconstruct the x of class y, which is most likely under the latent variable z. This encoder-decoder setup is realized using a neural network as shown in Fig. 5.1, and known as conditional VAE (CVAE). Once CVAE is properly trained, the decoder can be used to generate samples of any particular class, say y, by sampling z from a standard normal Gaussian, concatenating it with the respective a and passing it to the decoder.

Since VAE is a generative model, it can generate synthetic samples based on the learned distribution. However, it cannot perform conditional generation, i.e., generate samples for some specific class. For this purpose, a conditional VAE (Sohn et al., 2015) has been developed and employed for ZSL (Mishra et al., 2018). CVAE maximizes the variational lower bound as follows:

$$\mathcal{L}_{\text{CVAE}} = -\text{KL}(q_\Phi(z|x, c) || p_\theta(z|c)) + \mathbb{E}_{q_\Phi(z|c)} [\log p_\theta(x|z, c)], \quad (5.5)$$

where c is the condition. Here, the class-specific attribute information a is used as the condition. It allows us to generate samples based on the attributes a. It is noted that one wants $q(z)$ to be close to the standard normal distribution. Hence, $p_\theta(z|c)$ is taken as $N(0, I)$.

Based on CVAE, we present the proposed generative model-based CZSL method in the following section.

5.3.3 *Generative replay-based CZSL*

In this chapter, we propose a method for the single-head setting, i.e., when task information is not available at the inference time in the testing data stream (i.e., no task boundary). For this purpose, a GRCZSL method is proposed in this chapter. A flow diagram of the GRCZSL method is depicted in Fig. 5.2. The GRCZSL mainly has two components: (i) generative method for performing generalized ZSL, and (ii) generative replay for performing continual learning with

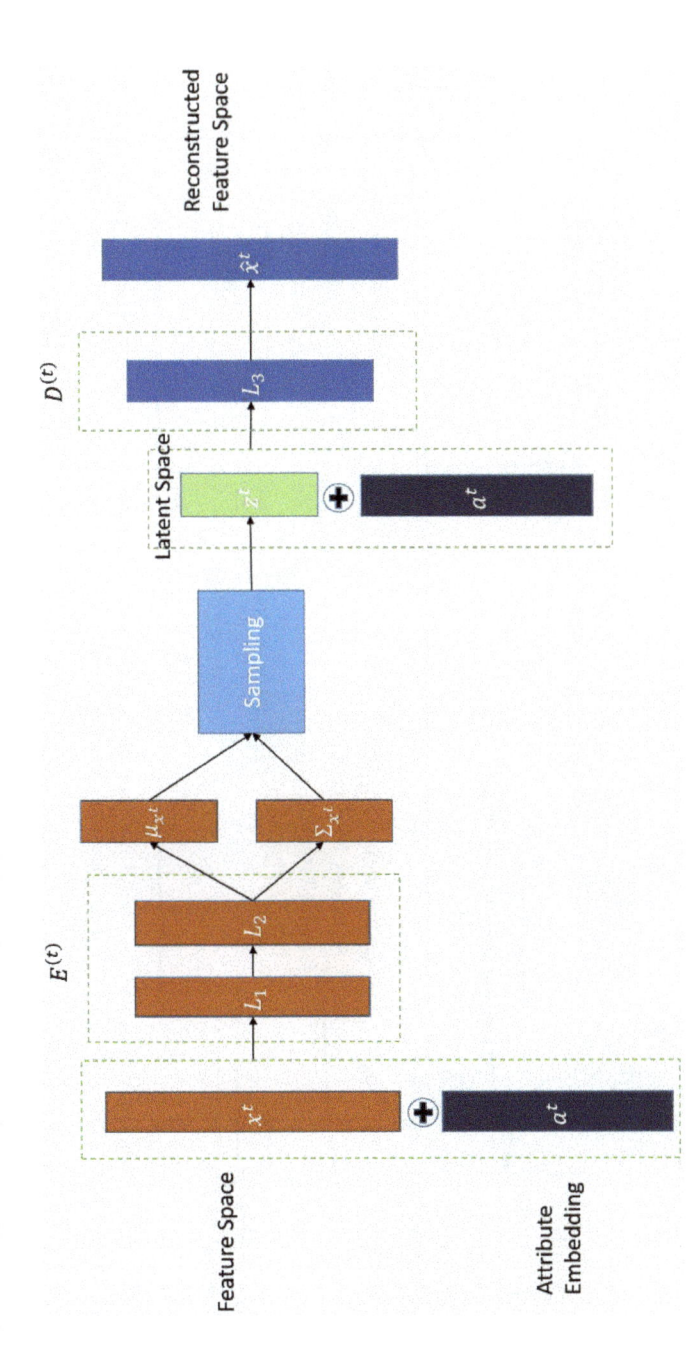

Fig. 5.1: Network architecture of CVAE used in the proposed GRCZSL method: The input feature x and the respective attribute vector a_y are concatenated and passed through a fully connected layer (L_1) of 512 units, followed by dropout with a dropout rate of 0.3 and passed through the fully connected layer (L_2) of 512 units. From (L_2), μ_z and Σ_z are further obtained individually via another fully connected layer of dimension 50. z is sampled from the variational distribution $\mathcal{N}(\mu_x, \Sigma_x)$. The sampled z concatenated with a_y is passed to a fully connected layer (L_3) of 1,024 units and the original x is then reconstructed. In this architecture, we use ReLU as an activation function at all layers, except the outputs of encoder and decoder, where we use a linear activation.

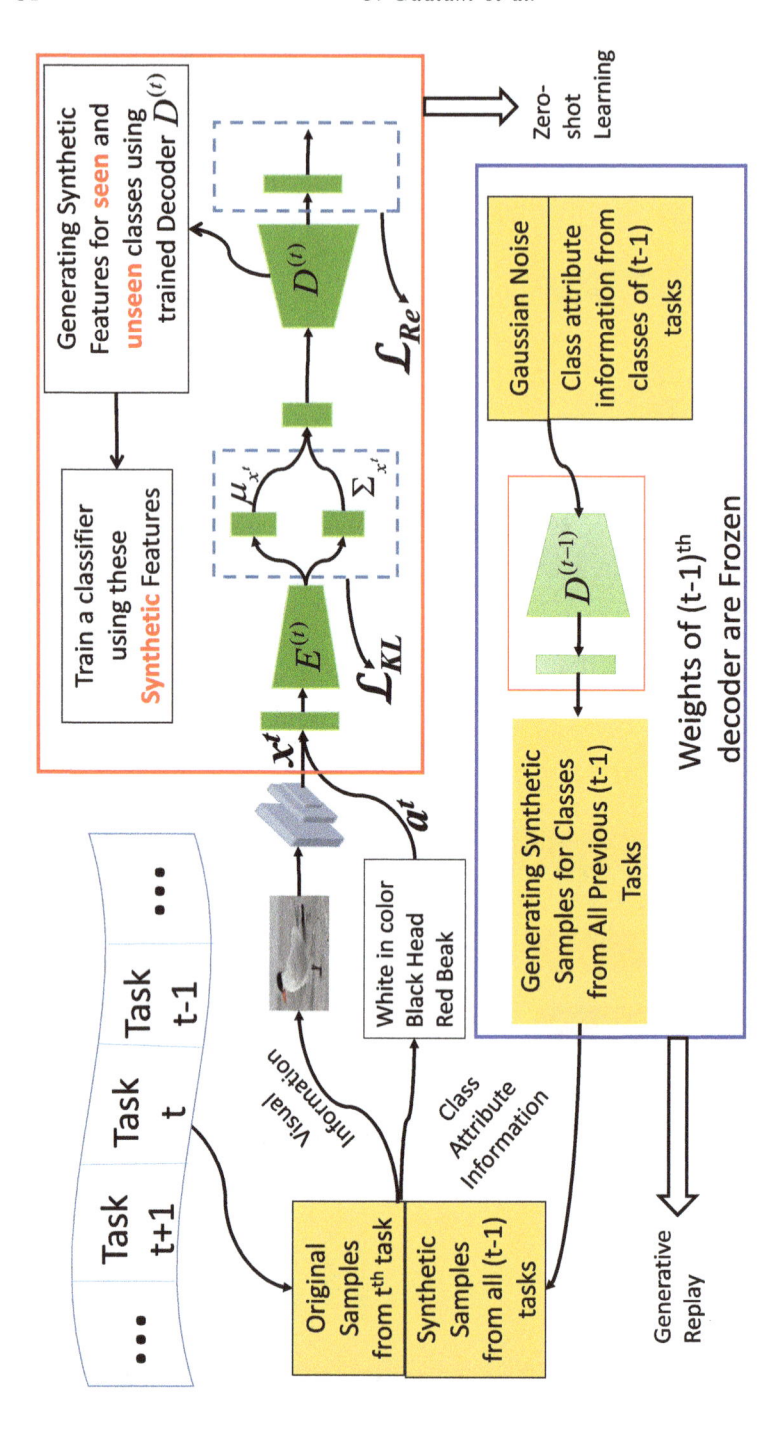

Fig. 5.2: Flow diagram of GRCZSL: $E^{(t)}$ and $D^{(t)}$ are the encoder and decoder for the tth task, which learns from the replay data and also performs CZSL. $D^{(t-1)}$ is the decoder of the $(t-1)$th task, which generates sample for generative replay.

ZSL. A standard generative method is used in both components, and CVAE is adapted for this purpose. However, one can use any generative model (the architecture of CVAE is detailed in Section 5.3.2). GRCZSL first performs a traditional ZSL on the first task and employs CZSL from the second task onward.

For the first task: GRCZSL is performed as follows:

(1) Train the CVAE for the training samples of the seen classes in the first task. A CVAE (Sohn et al., 2015) consists of an encoder (E) and decoder (D) network with parameters Φ and θ, respectively. It is a graphical model which estimates the mean ($\mu_{x_i^1}$) and variance ($\Sigma_{x_i^1}$) from the input stream x_i^1 using an encoder, i.e., $q_\Phi(z_i^1|x_i^1, a_i^1)$. Here, z_i^1 represents a latent variable, which is sampled from the estimated $\mathcal{N}(\mu_{x_i^1}, \Sigma_{x_i^1})$. Further, the latent variable z_i^1 is concatenated with the class attribute a_i^1 and passed to the decoder D so as to try to reconstruct the input x_i^1 at the output of the decoder. Overall, CVAE in the first task minimizes two losses simultaneously: KL divergence (Kullback and Leibler, 1951) loss ($\mathcal{L}_{\text{KL}}$) at the latent space and reconstruction loss ($\mathcal{L}_{\text{Re}}$) at the output of the decoder of CVAE, as follows:

$$\mathcal{L}(\theta^1, \Phi^1; x^1, a^1) = \mathcal{L}_{\text{Re}}^1(x^1, \hat{x}^1) + \mathcal{L}_{\text{KL}}^1\left(\mathcal{N}(\mu_{x^1}, \Sigma_{x^1}), \mathcal{N}(0, I)\right), \quad (5.6)$$

where $\hat{x}^1$ is the predicted value by CVAE. For reconstruction loss, the L_2 norm has been used.

(2) After training the CVAE model, synthetic samples for the seen and unseen classes are generated using the decoder of the trained CVAE model. For this, Gaussian noise with a concatenation of seen and unseen classes attribute information is passed to the trained decoder. This trained decoder generates synthetic samples for each seen and unseen class as class attribute information.

(3) After generating synthetic samples for the seen and unseen classes, a classifier is trained with these synthetic samples.

For all tasks except the first: After training the first task, the trained network from the previous task $t-1$ is used to train for the new task t as follows:

(1) For any tth task, the weights of the $(t-1)$th task's CVAE are frozen after the training.

(2) Now, the tth task's CVAE is initialized by the final weights of the previous $(t-1)$th task's CVAE and then trained.
(3) For training the tth task's CVAE, we pass the samples of seen classes at the tth task and the synthetic samples of the seen classes (x') of all previous $(t-1)$ tasks, which are generated before the training of the tth CVAE. To generate these synthetic samples (x') of the classes of previous tasks, GRCZSL uses the trained decoder of the $(t-1)$th task. These synthetic samples of the previous tasks helps GRCZSL in alleviating catastrophic forgetting of the CZSL.
(4) Overall, the tth task's CVAE minimizes the same two kinds of losses as the first task (i.e., reconstruction loss $(\mathcal{L}_{\text{Re}}^t)$ and KL divergence loss $(\mathcal{L}_{\text{KL}}^t)$ as follows:

$$\mathcal{L}(\theta^t, \Phi^t; x^t, a^t) = \alpha(\mathcal{L}_{\text{Re}}^t(x^t, \hat{x}^t) + \mathcal{L}_{\text{KL}}^t\left(\mathcal{N}(\mu_{x^t}, \Sigma_{x^t}), \mathcal{N}(0, I)\right)) \\ + (1-\alpha)(\mathcal{L}_{\text{Re}}^t(x', \hat{x}') \\ + \mathcal{L}_{\text{KL}}^t(\mathcal{N}(\mu_{x'}, \Sigma_{x'}), \mathcal{N}(0, I))), \qquad (5.7)$$

where α denotes task importance, μ_{x^t} and Σ_{x^t} represent the estimated mean and variance, respectively, for the tth task using the tth CVAE and $\hat{x}'$ is the corresponding predicted value of x' by the tth CVAE. Here, tth CVAE denotes the trained CVAE model after the tth task.
(5) After training the CVAE for the tth task, we generate synthetic samples for seen and unseen classes by the decoder of the tth CVAE, like in the first task. At the end, the classifier is trained for the tth task using these synthetic samples (a more detailed discussion is available in Section 5.3.3.1).

5.3.3.1 *Class-incremental learning in GRCZSL*

As discussed earlier, the CZSL problem is solved in a single-head setting, and it does not require task identity at the prediction time. If all training data are available *a priori* for ZSL, then we can jointly train over all classes using any ZSL method. However, in CZSL, all samples appear sequentially, and these samples collectively arrive as a sequence of tasks. Using the generative reply approach, GRCZSL generates synthetic features using the tth decoder of GRCZSL for the desired class of t tasks and uses these synthetic features for classification. A generated synthetic feature represents a synthetic sample of a specific class. In this way, GRCZSL can generate the desired number of samples for any underrepresented classes, which helps reduce the biases due to class imbalance. A single-hidden layer-based nonlinear classifier using a softmax function is used as a classifier. Here, the classifier continually learns new classes as the next tasks are presented to the model.

Overall, a CZSL method, namely GRCZSL, has been proposed. It requires only two networks current (model at the tth task) and previous (model at the $(t-1)$th task) models at any point in time during the training of a CZSL model. It is to be noted that only a decoder is required from the previous model, as a decoder is sufficient to generate synthetic samples for previous tasks. These factors make the GRCZSL a memory-efficient method.

Once the CVAE is trained on the tth task, the classifier is trained on the synthetic samples of all the classes up to task t. During inference, the trained classifier is used to classify the test data $(D_{\tau s}^t)$.

5.4 Performance Evaluation

In this section, we evaluate the performance of GRCZSL using five ZSL benchmark datasets, namely Attribute Pascal and Yahoo (aPY) (Farhadi et al., 2009), Animals with Attributes (AWA1 and AWA2) (Farhadi et al., 2009), Caltech-UCSD-Birds 200-2011 (CUB) (Wah et al., 2011) and SUN (Patterson and Hays, 2012). The data are split into various tasks, and the details are given in Table 5.2. For the purpose of performance comparison, we have used one of the

Table 5.2: Standard split of ZSL datasets.

Dataset	Attribute dimension	Seen classes	Unseen classes	Total classes
CUB	312	150	50	200
aPY	64	20	12	32
AWA1	85	40	10	50
AWA1	85	40	10	50
SUN	102	645	72	717

Table 5.3: Hyperparameters for GRCZSL in fixed CZSL setting.

Parameters	aPY	AWA1	AWA2	CUB	SUN
Learning rate (VAE)	0.001	0.001	0.001	0.001	0.001
Batch size (VAE)	50	50	50	50	50
Samples generated per seen class (VAE)	125	200	200	50	50
Batch size of generated samples (VAE)	15	20	50	100	100
Training epochs (VAE)	25	25	25	25	25
Hidden neurons (classifier)	1024	1024	1024	1024	512
Learning rate (classifier)	0.0001	0.0001	0.0001	0.0001	0.0001
Weight decay (classifier)	0.001	0.001	0.001	0.001	0.001
Batch size (classifier)	100	100	100	100	100
Training epochs (classifier)	30	30	30	10	25

Table 5.4: Hyperparameters for GRCZSL in dynamic CZSL setting.

Parameters	aPY	AWA1	AWA2	CUB	SUN
Learning rate (VAE)	0.001	0.001	0.001	0.001	0.001
Batch size (VAE)	50	50	50	50	50
Samples generated per seen class (VAE)	125	200	250	50	50
Batch size of generated samples (VAE)	15	800	20	100	100
Training epochs (VAE)	25	25	25	25	25
Hidden neurons (classifier)	1024	1024	1024	1024	512
Learning rate (classifier)	0.0001	0.0001	0.0001	0.0001	0.0001
Weight decay (classifier)	0.001	0.001	0.001	0.001	0.001
Batch size (classifier)	100	100	100	100	100
Training epochs (classifier)	30	30	30	10	25

standard settings and compared it with the existing state-of-the-art CZSL. Further, we have proposed another realistic setting to evaluate GRCZSL and compared it with the baseline. The respective performance metrics for different settings are described in the respective settings. All optimal hyperparameters are provided in Tables 5.3 and 5.4 for different settings. These settings and evaluation metrics are discussed in the subsequent section.

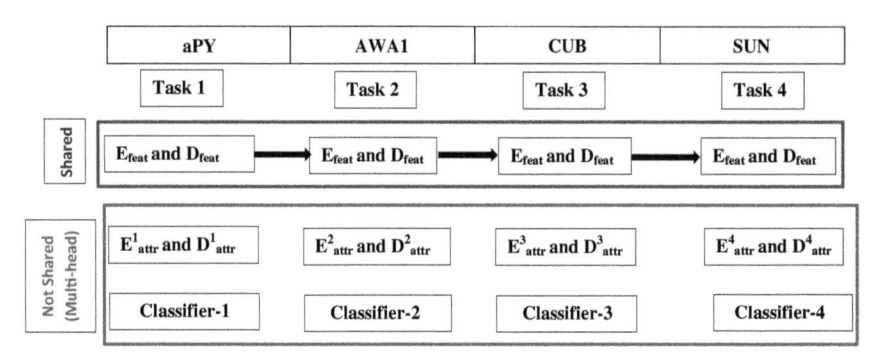

Fig. 5.3: CZSL setting by Wei et al. (2020).

5.4.1 Experimental settings and evaluation metrics

In the literature, two kinds of CZSL settings exist (Wei et al., 2020; Skorokhodov and Elhoseiny, 2021). The setting proposed by Wei et al. (2020) is a multi-head setting, as shown in Fig. 5.3, which is infeasible in real time. As it can be observed in Fig. 5.3, each task is considered a separate dataset, and even a distinct classifier and attribute-based distinct encoder-decoder for each task are deployed in the developed CZSL model, which is not feasible in real time. Furthermore, it is not a good solution for the multi-head setting.

Another setting is proposed by Skorokhodov and Elhoseiny (2021), which is a fixed CZSL setting, in which the number of classes is fixed at the very first task. In this section, we first discuss this fixed CZSL setting and its limitations and then discuss the new CZSL setting, i.e., the dynamic CZSL setting. Both experimental settings (fixed and dynamic) are designed based on the assumption of seen and unseen classes for each task. The details of each setting are provided as follows.

Fixed CZSL Setting: In this setting, all classes up to the current task are assumed to be the seen classes, and the classes from all remaining tasks are assumed to be the unseen classes, as shown in Fig. 5.4(a). This setting is identical to the setting presented by Skorokhodov and Elhoseiny (2021). The following evaluation metrics are used for the performance analysis of this setting:

Total no. of Classes in aPY = 32		Task-1	Task-2	Task-3	Task-4
Training Task Stream	Seen Classes	C1-C8	C9-C16	C17-C24	C25-C32
Testing Task Stream	Seen Classes	C1-C8	C1-C16	C1-C24	C1-C32
	Unseen Classes	C9-C32	C17-C32	C25-C32	--

(a) Fixed CZSL setting (Skorokhodov and Elhoseiny, 2021)

Total no. of Classes in aPY = 32		Task-1	Task-2	Task-3	Task-4
Training Task Stream	Seen Classes	C1-C5	C6-C10	C11-C15	C16-C20
Testing Task Stream	Seen Classes	C1-C5	C1-C10	C1-C15	C1-C20
	Unseen Classes	C21-C23	C21-C26	C21-C29	C21-C32

(b) Dynamic CZSL setting

Fig. 5.4: Fixed and dynamic CZSL settings.

- Mean seen-class accuracy (mSA):

$$\text{mSA} = \frac{1}{T}\sum_{t=1}^{T} CAcc(\mathcal{D}_{Ts}^{\leq t}, A^{\leq t}), \quad (5.8)$$

where the $CAcc$ function computes per class accuracy, T denotes the total number of tasks, $\mathcal{D}_{Ts}$ denotes testing data, $\mathcal{D}^{\leq t}$ denotes all train/test data from the first to the tth task and $A^{\leq t}$ is the corresponding class attribute of those unseen classes.

- Mean unseen-class accuracy (mUA):

$$\text{mUA} = \frac{1}{T-1}\sum_{t=1}^{T-1} CAcc(\mathcal{D}_{Ts}^{> t}, A^{> t}), \quad (5.9)$$

where $\mathcal{D}^{>t}$ denotes all train/test data from the $(t+1)$th to the last task and $A^{>t}$ is the corresponding class attribute of those unseen classes.

- Mean harmonic accuracy (mH):

$$\text{mH} = \frac{1}{T-1}\sum_{t=1}^{T-1} H(\mathcal{D}_{Ts}^{\leq t}, \mathcal{D}_{Ts}^{> t}, A), \quad (5.10)$$

where H stands for harmonic mean, $\mathcal{D}^{\leq t}$ denotes all train/test data from the first to the tth task, $\mathcal{D}^{>t}$ denotes all train/test data from the $(t+1)$th to the last task and A denotes the set of all class attributes.

Limitation of Fixed CZSL Setting: Since all classes from all tasks are available as seen or unseen classes, the setting cannot be utilized for a class-incremental and task-free learning setup of CZSL. Note that it is an infeasible assumption that all classes' attribute information is known in the first task.

Dynamic CZSL Setting: The new setting is proposed for two reasons: (i) testing the method for a class-incremental setup (i.e., dynamical), and (ii) splitting the data such that the standard seen–unseen classes' split of the ZSL dataset (Xian et al., 2017) is maintained at the last task's seen–unseen classes' split of the CZSL dataset. In this setting, the seen/unseen classes of the standard split are divided among various tasks for the CZSL experiment, as shown in Fig. 5.4(b). If any class is treated as seen/unseen for the tth task, those classes will be seen/unseen for all the subsequent tasks. During the evaluation of the model at the tth task, testing is performed on the current and all previous tasks' test data (but not on any of the next tasks as in the fixed CZSL setting). Similar to standard ZSL evaluation, the test data of the tth task in CZSL consists of 20% data from seen classes of the tth task and unseen classes of the tth task. If one follows the above-mentioned setting, the testing data of the last task in CZSL (continual learning) is identical to the testing data of the standard split in ZSL (offline learning). The following evaluation metrics are used for the performance analysis of this setting:

- Mean seen accuracy (mSA):

$$\text{mSA} = \frac{1}{T}\sum_{t=1}^{T} CAcc(\mathcal{D}_{Ts_s}^{\leq t}, A_{Ts_s}^{\leq t}), \tag{5.11}$$

where T denotes the total number of tasks, $\mathcal{D}_{Ts_s}^{t}$ denotes the test data of the tth task from seen classes and $A_{Ts_s}^{\leq t}$ is the corresponding class attribute of those seen classes.

- Mean unseen accuracy (mUA):

$$\text{mUA} = \frac{1}{T}\sum_{t=1}^{T} CAcc(\mathcal{D}_{Ts_{\text{us}}}^{\leq t}, A_{Ts_{\text{us}}}^{\leq t}), \tag{5.12}$$

where $\mathcal{D}_{Ts_{\text{us}}}^{t}$ denotes the unseen test data of the tth task and $A_{Ts_{\text{us}}}^{\leq t}$ is the corresponding class attribute of those unseen classes.

- Mean harmonic accuracy (mH):

$$\text{mH} = \frac{1}{T}\sum_{t=1}^{T} H(\mathcal{D}_{Ts_s}^{\leq t}, \mathcal{D}_{Ts_{\text{us}}}^{\leq t}, A^{\leq t}), \qquad (5.13)$$

where T denotes the total number of tasks, $\mathcal{D}_{Ts_s}^{t}$ denotes the test data of the tth task from seen classes, $\mathcal{D}_{Ts_{\text{us}}}^{t}$ denotes the unseen test data of the tth task and A denotes the set of all class attributes.

It is to be noted that both settings of CZSL are evaluated for generalized ZSL, i.e., GCZSL. A description of all datasets' split for the fixed and dynamic CZSL settings is provided in the subsequent section.

5.4.2 Dataset division as per fixed and dynamic settings for CZSL

As mentioned in the above section, two kinds of settings are used for the experiments. The datasets split as per both settings are mentioned as follows.

For Fixed CZSL Setting: The 200 classes of the CUB dataset are split into 20 tasks of 10 classes each. Similarly, the aPY dataset, which contains 32 classes, is split into 8 tasks with 4 classes each. The AWA1 and AWA2 datasets, which have 50 classes each, are split into 10 tasks with 5 classes each. The SUN dataset has 717 classes and is difficult to split evenly. Hence, it is split into 15 tasks, with 47 classes in the first 3 tasks and 48 classes in the remaining tasks. For all datasets, 20% of the data from each task are taken as test data to compute the final evaluation metrics.

For Dynamic CZSL Setting: The CUB dataset is split into 20 tasks, with the first 10 tasks containing 7 seen classes and 3 unseen classes each and the next 10 tasks containing 8 seen classes and 2 unseen classes each. Here, the test data consist of the unseen classes and 20% data from seen classes of each task. The aPY dataset is split into 8 tasks, with the first 4 tasks containing 2 seen classes and 2 unseen classes each and the remaining tasks containing 3 seen classes and 1 unseen classes each. The AWA1 and AWA2 are split into 10 tasks, with 4 seen classes and 1 unseen class per task. The

SUN dataset is split into 15 tasks, with the first 3 tasks containing 43 seen classes and 4 unseen classes each and the remaining tasks containing 43 seen classes and 5 unseen classes each.

5.4.3 Comparative analysis

There are only a handful of research works available on CZSL. We found three works so far, which can be categorized under multi-head (Chaudhry et al., 2018b; Wei et al., 2020) and single-head settings (Skorokhodov and Elhoseiny, 2021). Since the single-head setting is more feasible for practical scenarios (Chaudhry et al., 2018a), the single-head setting is used for experimental evaluation. The single-head setting-based method (Skorokhodov and Elhoseiny, 2021) has been used for a fair comparison. The following baselines are considered for a comparative study:

- **AGEM + CZSL (Skorokhodov and Elhoseiny, 2021):** Skorokhodov and Elhoseiny (2021) developed various CZSL methods based on average gradient episodic memory (Chaudhry et al., 2018b), elastic weight consolidation (Kirkpatrick et al., 2017; Schwarz et al., 2018) and memory aware synapses (Aljundi et al., 2018) for single-head setting using a fixed CZSL setting. Since the authors presented results on only two datasets, namely the CUB and SUN datasets, the results on these two datasets are provided in Table 5.5 for comparison.
- **Sequential Baselines:** For developing baselines for CZSL, we sequentially train the base classifier with CVAE over multiple tasks without considering any continual learning strategy. In sequential training, the initial weight of the CVAE at the tth task is the final weight of CVAE at the $(t-1)$th task. After training the Seq-CVAE on the current task, synthetic samples are generated using the trained decoder (D_t), as well as the class attribute information for all the classes which need to be classified. Here, two types of regularizations (L1 and L2) have been utilized to generate two baseline methods (SeqL1 and SeqL2, respectively). For creating this baseline and in the proposed method, CVAE is used as the base autoencoder as it generates appropriate feature space for different classes as shown in Fig. 5.5.

Table 5.5: mSA, mUA and mH values are provided for fixed CZSL setting. The bold face represents the best results in the table.

	CUB mSA	CUB mUA	CUB mH	aPY mSA	aPY mUA	aPY mH	AWA1 mSA	AWA1 mUA	AWA1 mH	AWA2 mSA	AWA2 mUA	AWA2 mH	SUN mSA	SUN mUA	SUN mH
SeqL1	15.25	5.33	7.75	28.81	5.19	8.07	38.52	8.59	13.05	40.25	9.67	14.04	7.41	3.92	5.04
SeqL2	24.66	8.57	12.18	46.99	13.04	19.12	44.52	15.03	21.67	47.94	16.02	23.17	16.88	11.40	13.38
AGEM+CZSL (Skorokhodov and Elhoseiny, 2021; Chaudhry et al., 2018b)	—	—	17.30	—	—	—	—	—	—	—	—	—	—	—	9.60
EWC+CZSL (Skorokhodov and Elhoseiny, 2021; Schwarz et al., 2018)	—	—	18.00	—	—	—	—	—	—	—	—	—	—	—	9.60
MAS+CZSL (Skorokhodov and Elhoseiny, 2021; Aljundi et al., 2018)	—	—	17.70	—	—	—	—	—	—	—	—	—	—	—	9.40
GRCZSL	**41.91**	**14.12**	**20.48**	**62.27**	**12.57**	**20.46**	**77.36**	**23.24**	**34.86**	**80.57**	**24.35**	**36.57**	**17.74**	**11.50**	**13.73**

(a) ResNet-101 features   (b) Synthetic features generated using CVAE

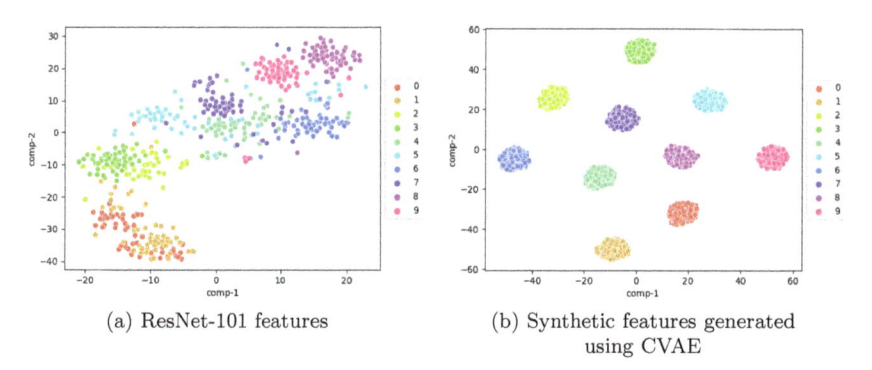

Fig. 5.5: *t*-SNE plot of 10 random classes of the CUB dataset for ResNet-101 features and synthetic features generated through CVAE.

Extensive experiments have been performed based on both settings for GRCZSL, and performance analysis is discussed as follows.

GRCZSL for Fixed Setting: Table 5.5 contains results for the fixed CZSL setting over five datasets. For the CUB dataset, GRCZSL significantly outperforms the two baselines by more than 8% in terms of mH. It yields more than 3%, 2% and 2% mH values compared to the state-of-the-art CZSL methods, AGEM + CZSL (Skorokhodov and Elhoseiny, 2021), EWC + CZSL (Skorokhodov and Elhoseiny, 2021; Kirkpatrick *et al.*, 2017) and MAS + CZSL (Skorokhodov and Elhoseiny, 2021; Aljundi *et al.*, 2018), respectively. Similarly, GRCZSL improves the performance over the baselines by more than 11% and 12% for the AWA1 and AWA2 datasets, respectively. These improvements have been gained due to the replay of the synthetic samples generated by the immediately previous model. Here, we do not need to store any samples from the real data explicitly. At any point in time during training for the *t*th task, GRCZSL requires only two models, i.e., one for the current task and the other for the previous task. The proposed method exhibits the least mH improvements over baselines on the aPY and SUN datasets by more than 1.5% and 0.3%, respectively.

GRCZSL for Dynamic Setting: Table 5.6 contains results for the fixed CZSL setting over five datasets. GRCZSL outperforms all baseline methods for all datasets. It exhibits a significant improvement of more than 8%, 6%, 6%, 6% and 1% for the CUB, aPY, AWA1, AWA2 and SUN datasets, respectively. As mentioned above,

Table 5.6: mSA, mUA and mH values are calculated for dynamic CZSL setting. The bold face represents the best results in the table.

	CUB			aPY			AWA1			AWA2			SUN		
	mSA	mUA	mH	mSA	mUA	mH	mSA	mUA	mH	mSA	mUA	mH	mSA	mUA	mH
SeqL1	25.71	12.78	16.76	52.18	6.93	11.24	49.90	15.32	22.50	55.46	14.16	21.98	15.36	9.28	11.03
SeqL2	38.95	20.89	26.74	61.12	14.22	22.91	63.11	30.98	40.84	67.68	35.64	46.01	29.06	21.33	24.33
GRCZSL	**59.27**	**26.03**	**35.67**	**77.14**	**17.94**	**28.75**	**86.92**	**33.21**	**47.48**	**90.61**	**37.56**	**52.61**	**30.78**	**22.59**	**25.54**

Table 5.7: The table provides results of the CZSL methods at the last task when the train–test split is identical to the offline case, where all training and testing data are available at once. We also provide the results of the offline method, which is basically an upper bound for the CZSL methods available in the table.

	CUB	aPY	AWA1	AWA2	SUN
Offline (upper bound)	34.50	22.38	47.20	51.20	26.7
SeqL1	12.17	0.72	20.70	16.26	8.21
SeqL2	14.89	14.11	16.29	25.36	16.85
GRCZSL	**26.34**	**17.68**	**37.06**	**38.71**	**20.80**

for this setting, the performance on the last task needs to be equivalent to the base offline method CVAE, which is the upper bound for our proposed method. The mH values on the last task for all methods are presented in Table 5.7. It can be analyzed from this table that GRCZSL exhibits better performance than baselines; however, there is still room for improvement, as GRCZSL falls behind the upper bound.

5.4.4 Ablation study on the CUB dataset

The ablation study is conducted using the CUB dataset.

Task-wise Performance Analysis: Task-wise results are plotted in Fig. 5.6 in terms of mH. These plots are generated by providing equal importance to the samples from the current and previous tasks. We observed that overall performance (i.e., mH) and performance on unseen classes (i.e., mUA) follow a similar pattern over 20 tasks of the CUB dataset for each setting (please refer to Fig. 5.7 for the plot of mUA). We observe this pattern due to the lower performance of unseen classes, which is the main reason for the lower performance of any model in ZSL. However, the fixed and dynamic settings exhibit different patterns in Fig. 5.6 because the number of unseen classes is decreasing and the number of seen classes increases with each task in the case of the fixed CZSL setting. In contrast, the numbers of seen and unseen classes are constantly increasing with each task in the case of the dynamic CZSL setting. Therefore, we observed different patterns in both settings.

(a) For fixed CZSL setting
(b) For dynamic CZSL setting

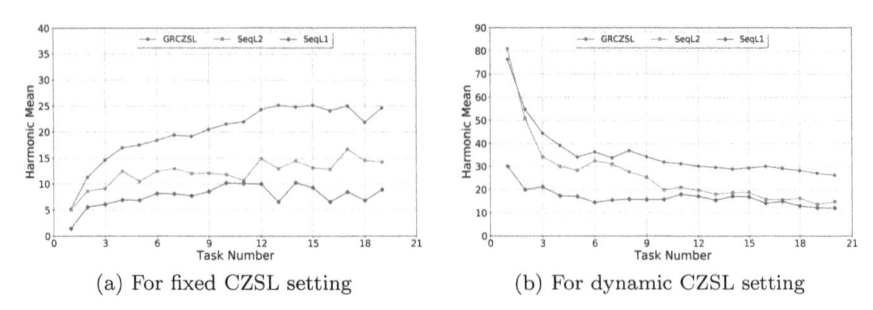

Fig. 5.6: Harmonic mean for the CUB dataset over 20 tasks for fixed and dynamic CZSL settings.

(a) For fixed CZSL setting
(b) For dynamic CZSL setting

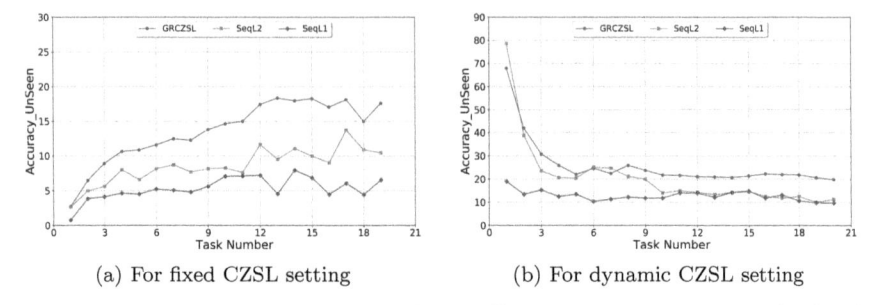

Fig. 5.7: Per class unseen accuracy for the CUB dataset over 20 tasks for fixed and dynamic CZSL settings.

Task Importance Analysis: To analyze the importance of the current and previous tasks, task importance (α) is varied in the range $\alpha = [0, 0.1, 0.3, 0.5, 0.7, 0.9, 1]$. Similarly to the above discussion, the same pattern is observed between the overall performance and the performance on unseen classes over 20 tasks of the CUB dataset (see Fig. 5.8 for the plots). It can be analyzed from Fig. 5.9 that GRCZSL performed worst when $\alpha = 0$, i.e., there is no importance of the current task. The second-worst performance is observed when $\alpha = 1$, i.e., there is no importance of previous tasks. The best performance is observed when task importance is in the range of $[0.3, 0.5]$ for both settings.

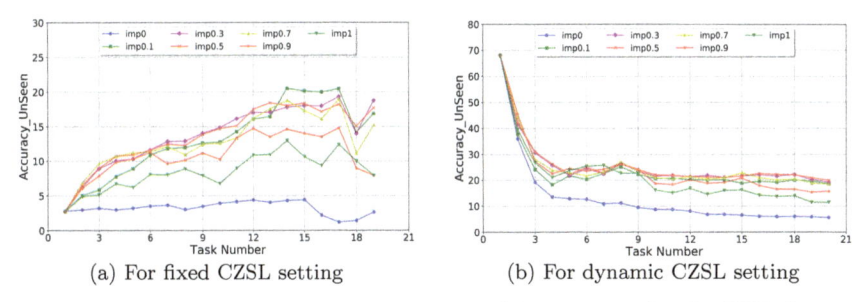

Fig. 5.8: Per class unseen accuracy as per task importance for the CUB dataset over 20 tasks for fixed and dynamic CZSL settings.

(a) For fixed CZSL setting

(b) For dynamic CZSL setting

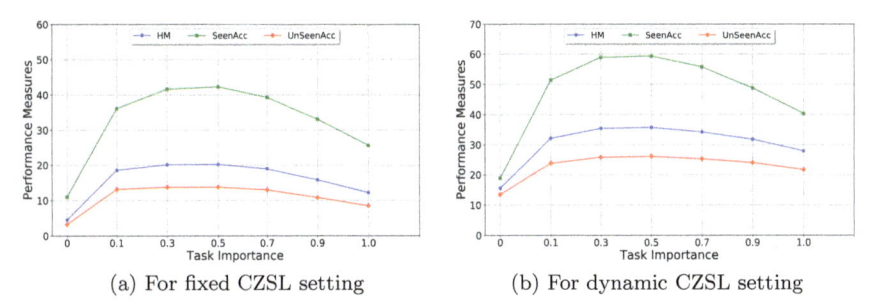

Fig. 5.9: The impact of the task importance on the CUB dataset in terms of mH, mSA and mUA for both settings.

5.5 Conclusion

In this chapter, we presented a GRCZSL method for a single-head setting, capable of handling unseen classes by only using semantic information. The GRCZSL only requires an entire network for the current task and a previously trained decoder alone from the immediate previous task. The network performs CZSL, and the previously trained (only decoder) network alleviates catastrophic forgetting by generating replay samples for the classes of the previously learned tasks. These samples help the network retain the knowledge of the earlier tasks. GRCZSL also provides a balance between the former and the new task's performance. Overall, the proposed GRCZSL can be used in the continual/lifelong learning setting without retraining from scratch. The experimental results on five datasets and the ablation study on the CUB dataset also exhibited that GRCZSL significantly outperformed the baseline and the existing state-of-the-art methods. The current CZSL evaluation setting problem is alleviated in this study through new CZSL class-incremental and task-free learning settings. The performance of GRCZSL has been evaluated in the setting, and it is compared with the baseline. Since GRCZSL requires task information during training, it is not ideal for task-free learning. Therefore, it is necessary to develop a CZSL method which will be appropriate for both class-incremental and task-free learning.

© 2024 World Scientific Publishing Company
https://doi.org/10.1142/9789811286711_0006

Chapter 6

Architect, Regularize and Replay: A Flexible Hybrid Approach for Continual Learning

Vincenzo Lomonaco[*,‡], Lorenzo Pellegrini[†,§],
Gabriele Graffieti[†,‖] and Davide Maltoni[†,¶]

[*]*University of Pisa, Pisa, Italy*
[†]*University of Bologna, Bologna, Italy*
[‡]*vincenzo.lomonaco@unipi.it*
[§]*l.pellegrini@unibo.it*
[‖]*gabriele.graffieti@unibo.it*
[¶]*davide.maltoni@unibo.it*

Abstract

In recent years, we have witnessed a renewed interest in machine learning methodologies, especially for deep representation learning, that could overcome basic i.i.d. assumptions and tackle non-stationary environments subject to various distributional shifts or sample selection biases. Within this context, several computational approaches based on architectural priors, regularizers and replay policies have been proposed with different degrees of success depending on the specific scenario in which they were developed and assessed. However, designing comprehensive hybrid solutions that can be flexibly and generally applied with tunable efficiency–effectiveness trade-offs still seems like a distant goal. In this chapter, we propose "Architect, Regularize and Replay" (ARR), a hybrid generalization of the renowned AR1 algorithm and its variants, which can not only achieve state-of-the-art results in classic scenarios (e.g., class-incremental learning) but also generalize to arbitrary data streams generated from real-world datasets, such as CIFAR-100, CORe50 and ImageNet-1000.

6.1 Introduction

Continual machine learning is a challenging research problem with profound scientific and engineering implications (Lomonaco, 2019). On the one hand, it undermines the foundations of classic machine learning systems relying on *i.i.d.* assumptions; on the other hand, it offers a path toward efficient and salable human-centered AI systems that can learn and think like humans, swiftly adapting to the ever-changing nature of the external world. However, despite the recent surge in interest from the machine learning and deep learning communities on the topic and the prolific scientific activity of the past few years, the issue is far from being solved.

While most continual learning (CL) algorithms significantly reduce the impact of catastrophic forgetting in specific scenarios, it is difficult to generalize those results to settings in which they have not been specifically designed to operate (lack of robustness and generality). Moreover, they are mostly focused on vertical and exclusive approaches to CL based on regularization, replay or architectural changes in the underlying prediction model.

In this chapter, we summarize the efforts made in the formulation of hybrid strategies for CL that can be more robust, generally applicable and effective in real-world application contexts. In particular, we focus on the definition of the *"Architect, Regularize and Replay"* (ARR) method: a general reformulation and generalization of the renowned AR1 algorithm (Maltoni and Lomonaco, 2019) with all its variants (Lomonaco *et al.*, 2019a; Pellegrini *et al.*, 2019) and, arguably, one of the first hybrid CL methods proposed (Parisi and Lomonaco, 2020). Through a number of experiments on state-of-the-art benchmarks, such as CIFAR-100, CORe50 and ImageNet-1000, we show the efficiency and effectiveness of the proposed approach with respect to other existing state-of-the-art methods.

6.2 Background and Problem Formulation

Continual learning is mostly concerned with the concept of learning from a stream of ephemeral, non-stationary data that can be processed in separate computational steps and cannot be revisited if not explicitly memorized. In an agnostic CL scenario, data arrive in a streaming fashion as a (possibly infinite) sequence S of, what we call, *learning experiences* e, so that $S = e_1, \ldots, e_n$. For simplicity, we assume a supervised classification problem, where each experience e_i consists of a batch of samples $\mathcal{D}^i$, with each sample being a tuple $\langle x_k^i, y_k^i \rangle$ of input and target data, respectively, and the labels y_k^i are from the set $\mathcal{Y}^i$, which is a subset of the entire universe of classes $\mathcal{Y}$. However, we note that this formulation is very easy to generalize to different CL problems. Usually, $\mathcal{D}^i$ is split into a separate train set $\mathcal{D}_{\text{train}}^i$ and test set $\mathcal{D}_{\text{test}}^i$. A CL algorithm $\mathcal{A}^{\text{CL}}$ is a function with the following signature (Lesort et al., 2019):

$$\mathcal{A}^{\text{CL}} : \langle f_{i-1}^{\text{CL}}, \mathcal{D}_{\text{train}}^i, \mathcal{M}_{i-1}, t_i \rangle \to \langle f_i^{\text{CL}}, \mathcal{M}_i \rangle \quad (6.1)$$

where f_i^{CL} is the model learned after training on experience e^i and $\mathcal{M}_i$ a buffer of past knowledge (can be also void), such as previous samples or activations, stored from the previous experiences and usually of fixed size. The term t_i is a task label which may be used to identify the correct data distribution (or *task*). All the experiments in this study assume the most challenging scenario of t_i being unavailable. Usually, CL algorithms are limited in the amount of resources that they can use, and they should be designed to scale up to a large number of training experiences without increasing their memory/computational overheads over time. The objective of a CL algorithm is to minimize the loss $\mathcal{L}_S$ over the entire stream of data S, composed of n distinct experiences:

$$\mathcal{L}_S(f_n^{\text{CL}}, n) = \frac{1}{\sum_{i=1}^n |\mathcal{D}_{\text{test}}^i|} \sum_{i=1}^n \mathcal{L}_{\exp}(f_n^{\text{CL}}, \mathcal{D}_{\text{test}}^i) \quad (6.2)$$

$$\mathcal{L}_{\exp}(f_n^{\text{CL}}, \mathcal{D}_{\text{test}}^i) = \sum_{j=1}^{|\mathcal{D}_{\text{test}}^i|} \mathcal{L}(f_n^{\text{CL}}(\boldsymbol{x}_j^i), y_j^i), \quad (6.3)$$

where the loss $\mathcal{L}(f_n^{\text{CL}}(\boldsymbol{x}), y)$ is computed on a single sample $\langle \boldsymbol{x}, y \rangle$, such as cross-entropy in classification problems. Hence, the main assumption in this formulation is that all the concepts encountered over time are still relevant (the drift is only *virtual*) and there is no conflicting evidence. This is quite a common assumption for the deep

CL literature, which is more concerned with building robust and general representations over time than building systems that can quickly adapt to changing circumstances.

6.3 Toward Hybrid Continual Learning Approaches

We show in Fig. 6.1 some of the most popular and recent CL approaches divided into the above-introduced categories and their combinations. In the diagram, we differentiate methods with *rehearsal* (replay of explicitly stored training samples) from methods with *generative replay* (replay of latent representations or training samples). Crucially, although an increasing number of methods have been proposed, there is no consensus on which training schemes and performance metrics are better to evaluate CL models. Different sets of metrics have been proposed to evaluate CL performance on supervised and unsupervised learning tasks (e.g., Hayes et al., 2018b; Kemker et al., 2017; Díaz-Rodríguez et al., 2018). In the absence of standardized metrics and evaluation schemes, it is unclear what

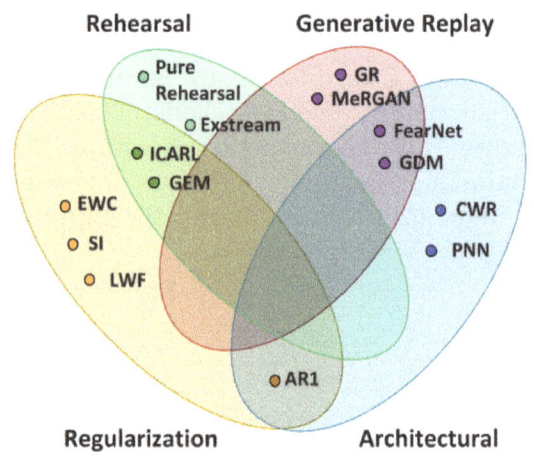

Fig. 6.1: A Venn diagram of some of the most popular CL strategies: CWR (Lomonaco and Maltoni, 2017), PNN (Rusu et al., 2016a), EWC (Kirkpatrick et al., 2017), SI (Zenke et al., 2017), LWF (Li and Hoiem, 2016), ICARL (Rebuffi et al., 2017b), GEM (Lopez-Paz and Ranzato, 2017), FearNet (Kemker and Kanan, 2018), GDM (Parisi et al., 2018), ExStream (Hayes et al., 2018a), Pure Rehearsal, GR (Shin et al., 2017), MeRGAN (Wu et al., 2018) and AR1 (Maltoni and Lomonaco, 2019). Rehearsal and generative replay upper categories can be seen as subsets of replay strategies.

it means to endow a method with CL capabilities. In particular, a number of CL models still require large computational and memory resources that hinder their ability to learn in real time, or with a reasonable latency, from data streams.

It is also worth noting that, while an overabundant proposal of methods falling into each of the main categories exists, it is still difficult to find hybrid algorithmic solutions that can flexibly leverage the often orthogonal advantages of the three different approaches (i.e., architectural, regularization and replay), depending on the specific application needs and target efficiency–effectiveness trade-off. However, some evidence exists that effective biological CL systems (such as the human brain) make use of all these distinct functionalities.

In this chapter, we argue that in the near- and long-term future of lifelong learning machines, *we will witness a significantly growing interest in the development of hybrid CL algorithms* (Lomonaco et al., 2022), and we propose ARR as one of the first methodologies that practically implements such a vision.

6.4 Architect, Regularize and Replay

The *architect*, *regularize* and *replay* algorithm is a flexible *generalization* of the AR1 algorithm and its variants (CWR+, CWR*, AR1* and AR1Free) (Lomonaco et al., 2020; Pellegrini et al., 2019). ARR, with a proper instantiation of its hyperparameters, can be instantiated in the aforementioned algorithms based on the desired efficiency–efficacy trade-off (Ravaglia et al., 2020). It can use pretrained parameters, as suggested by a consolidated trend in the field (Cossu et al., 2022b), or start from a random initialization. The pseudocode in Algorithm 6.1 describes the ARR in detail based on its three main components: *architectural*, *regularization* and *replay*.

6.4.1 *Architectural component*

The core concept behind an architectural approach is to *isolate* and preserve some parameters while adding new ones in order to house new knowledge. CWR+, an evolution of CWR (Lomonaco and Maltoni, 2017), whose pseudocode is reported in Algorithm 2 by Maltoni and Lomonaco (2019), maintains two sets of weights for the output classification layer: cw are the consolidated weights (for stability)

Algorithm 6.1 ARR pseudocode: $\bar{\Theta}$ are the class-shared parameters of the representation layers; the notation $cw[j]/tw[j]$ is used to denote the groups of consolidated/temporary weights corresponding to class j. Note that this version continues to work under new classes (NC), which is seen here as a special case of new classes and instances (NIC) (Lomonaco and Maltoni, 2017); in fact, since in NC the classes in the current batch were never encountered before, the step at line 5 loads a value of 0 for classes in B_i because cw_j were initialized to 0 and in the consolidation step (line 15) and $wpast_j$ values are always 0. The external random memory RM is populated across the training batches. Note that the number h of examples to add progressively decreases to maintain a nearly balanced contribution from the different training batches; however, no constraints are enforced to achieve class balancing. λ is the regularization strength, and α is the replay layer. The three input parameters will default to 0 if omitted.

1: Hyperparameters: RM_{size}, λ, α
2: $RM = \varnothing$, $cw[j] = 0$ and $past_j = 0$ $\forall j$
3: init $\bar{\Theta}$ randomly or from pretrained model (e.g., on ImageNet)
4: **for each** training batch B_i:
5: $\quad tw[j] = \begin{cases} cw[j], & \text{if class } j \text{ in } B_i \\ 0, & \text{otherwise} \end{cases}$
6: $\quad mb_e = \begin{cases} \frac{|B_i|}{(|B_i|+RM_{\text{size}})/mb_{\text{size}}}, & \text{if } B_i > B_1 \\ mb_{\text{size}}, & \text{otherwise} \end{cases}$
7: $\quad mb_r = mb_{\text{size}} - mb_e$
8: $\quad$ **for each** epoch:
9: $\quad\quad$ Sample mb_e examples from B_i and mb_r examples from RM
10: $\quad\quad$ train the model on sampled data (replay data to be injected at layer α):
11: $\quad\quad\quad$ **if** $B_i = B_1$ learn both $\bar{\Theta}$ and tw
12: $\quad\quad\quad$ **else** learn tw and $\bar{\Theta}$ with λ to control forgetting.
13: $\quad$ **for each** class j in B_i:
14: $\quad\quad wpast_j = \sqrt{\frac{past_j}{cur_j}}$, where cur_j is the number of examples of class j in B_i
15: $\quad\quad cw[j] = \frac{cw[j] \cdot wpast_j + (tw[j] - avg(tw))}{wpast_j + 1}$
16: $\quad\quad past_j = past_j + cur_j$
17: $\quad$ test the model using $\bar{\Theta}$ and cw
18: $\quad h = \frac{RM_{\text{size}}}{i}$
19: $\quad R_{\text{add}} = $ random sampling h examples from B_i
20: $\quad R_{\text{replace}} = \begin{cases} \varnothing & \text{if } i = 1 \\ \text{random sample } h \text{ examples from } RM & \text{otherwise} \end{cases}$
21: $\quad RM = (RM - R_{\text{replace}}) \cup R_{\text{add}}$

used for inference, and *tw* are the temporary weights (for plasticity) used for training; *cw* are initialized to 0 before the first batch and then iteratively updated, while *tw* are reset to 0 before each training batch.

Maltoni and Lomonaco (2019) proposed an extension of `CWR+` called `CWR*`, which works both under *class-incremental* (Rebuffi et al., 2017b) and *class-incremental with repetition* settings (Cossu et al., 2022a); in particular, under class-incremental with repetition, the coming batches include examples of both new and already encountered classes. For already known classes, instead of resetting weights to 0, consolidated weights are reloaded. Furthermore, in the consolidation step, a weighted sum is now used: The first term represents the weight of the past, and the second term is the contribution from the current training batch. The weight $wpast_j$ used for the first term is proportional to the ratio $\frac{past_j}{cur_j}$, where $past_j$ is the total number of examples of class j encountered in past batches, whereas cur_j is their count in the current batch. In the case of a large number of small non-i.i.d. training batches, the weights for the most recent batches may be too low, thus hindering the learning process. In order to avoid this, a square root is used to smooth the final value of $wpast_j$.

6.4.2 *Regularization component*

The well-known *elastic weight consolidation* (EWC) pure regularization approach (Kirkpatrick et al., 2017) controls forgetting by proportionally constraining the model weights based on their estimated importance with respect to previously encountered data distributions and tasks. To this purpose, in a classification approach, a regularization term is added to the conventional cross-entropy loss, where each weight θ_k of the model is pulled back to its optimal value θ_k^* with a strength F_k proportional to their estimated importance for modeling past knowledge:

$$L = L_{\text{cross}}(\cdot) + \frac{\lambda}{2} \cdot \sum_k F_k \cdot (\theta_k - \theta_k^*)^2. \tag{6.4}$$

Synaptic intelligence (SI) (Zenke et al., 2017) is an equally well-known lightweight variant of EWC, where, instead of updating the

Fisher information F at the end of each batch,[1] F_k are obtained by integrating the loss over the weight trajectories, exploiting information already available during gradient descent. For both approaches, the weight update rule corresponding to Eq. (6.4) is

$$\theta'_k = \theta_k - \eta \cdot \frac{\partial L_{\text{cross}}(\cdot)}{\partial \theta_k} - \eta \cdot F_k \cdot (\theta_k - \theta^*_k), \quad (6.5)$$

where η is the learning rate. This equation has two drawbacks. First, the value of λ must be carefully calibrated. In fact, if its value is too high, the optimal value of some parameters could overshoot, leading to divergence (see discussion by Maltoni and Lomonaco, 2019, Section 2). Second, two copies of all model weights must be maintained to store both θ_k and θ^*_k, leading to double memory consumption for each weight. To overcome the above problems, Lomonaco et al. (2020) proposed to replace the update rule of Eq. (6.5) with

$$\theta'_k = \theta_k - \eta \cdot \left(1 - \frac{F_k}{max_F}\right) \cdot \frac{\partial L_{\text{cross}}(\cdot)}{\partial \theta_k}, \quad (6.6)$$

where max_F is the maximum value for weight importance (we clip to max_F the F_k values larger than max_F). Basically, the learning rate is reduced to 0 (i.e., complete freezing) for weights of highest importance ($F_k = max_F$) and maintained at η for weights whose $F_k = 0$. It is worth noting that these two updated rules work differently: The former still moves weights with high F_k in a direction opposite to the gradient and then makes a step in the direction of the past (optimal) values; the latter tends to completely freeze weights with high F_k. However, in the experiments conducted by Lomonaco et al. (2020), the two approaches led to similar results; therefore, the second one is preferable since it overcomes the aforementioned drawbacks. The regularization of learning parameters can be enforced on both the low-level generic features and the class-specific discriminative features, as implemented in AR1*. However, for the sake of simplicity, in ARR, we consider only the application of such regularization terms to the last group since freezing or slowly fine-tuning the low-level generic features already proved to be an effective strategy.

---

[1] In this chapter, for the EWC and ARR implementations, we use a single Fisher matrix updated over time, following the approach described by Maltoni and Lomonaco (2019).

6.4.3 Replay component

Pellegrini *et al.* (2019) and Merlin *et al.* (2022) showed that a very simple rehearsal implementation (hereafter denoted as *native rehearsal*), where for every training batch a random subset of the batch examples is added to the external storage to replace an (equally random) subset of the external memory, is not less effective than more sophisticated approaches, such as iCaRL. Therefore, Pellegrini *et al.* (2019) opted for simplicity and compared the learning trend of CWR* and AR1* of a MobileNetV1[2] trained with and without rehearsal on CORe50 NICv2-391 (Lomonaco *et al.*, 2020). They used the same protocol and hyperparameters introduced by Lomonaco *et al.* (2019b) and a rehearsal memory of 1500 examples. It is well evident from their study that even a moderate amount of external memory (about 1.27% of the total training set) is very effective in improving the accuracy of both approaches and reducing the gap with the cumulative upper bound, which for this model was ~85%.

In deep neural networks, the layers close to the input (often denoted as representation layers) usually perform low-level feature extraction, and after proper pretraining on a large dataset (e.g., ImageNet), their weights are quite stable and reusable across applications. On the other hand, higher layers tend to extract class-specific discriminant features, and their tuning is often important to maximize accuracy.

A *latent replay* (see Fig. 6.2) approach (Pellegrini *et al.*, 2019) can then be formulated: Instead of maintaining copies of input examples in the external memory in the form of raw data, we can store the *activation volumes* at a given layer (denoted as the *latent replay layer*). To keep the representation stable and the stored activations valid, we propose to slow down the learning at all the layers below the latent replay one and to leave the layers above free to learn at full pace. In the limit case where lower layers are completely frozen (i.e., slow-down to 0), latent replay is functionally equivalent to rehearsal from the input but achieves computational and storage savings thanks to the smaller fraction of examples that need to flow forward and backward across the entire network and the typical information compression that networks perform at higher layers.

---

[2] The network was pretrained on ImageNet-1k.

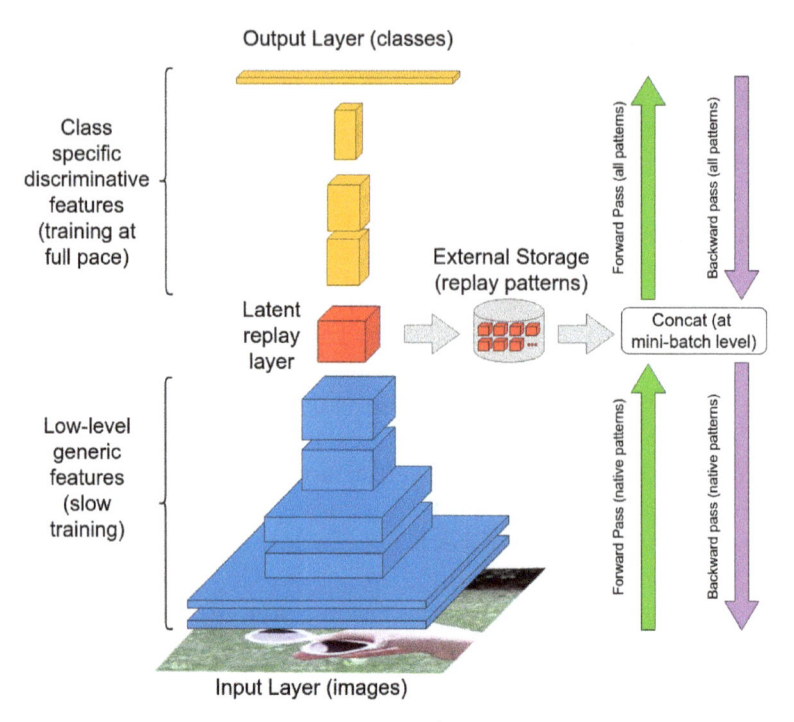

Fig. 6.2: Architectural diagram of ARR (Pellegrini *et al.*, 2019).

In the general case where the representation layers are not completely frozen, the activations stored in the external memory may suffer from an *aging effect* (i.e., as time passes, they tend to increasingly deviate from the activations that the same pattern would produce if feed-forward from the input layer). However, if the training of these layers is sufficiently slow, the aging effect is not disruptive since the external memory has enough time to be updated with newly acquired examples. When latent replay is implemented with mini-batch SGD training: (i) in the forward step, a concatenation is performed at the replay layer (on the mini-batch dimension) to join examples coming from the input layer with activations coming from the external storage; (ii) the backward step is stopped just before the replay layer for the replay examples.

6.5 Empirical Evaluation

In order to empirically evaluate the overall quality and flexibility of ARR, we evaluate its performance on three commonly used continual learning benchmarks for computer vision classification tasks: CIFAR-100 (Section 6.5.1), CORe50 (Section 6.5.2) and ImageNet-1000 (Section 6.5.3). Then, we provide a more in-depth analysis on the impact of the latent replay layer selection (Section 6.5.4) and the memory size in terms of memorized activation volumes (Section 6.5.5).

6.5.1 *CIFAR-100*

CIFAR-100 (Krizhevsky, 2009) is a well-known and widely used dataset for small (32 × 32) natural image classification. It includes 100 classes containing 600 images each (500 training + 100 test). The default classification benchmark can be translated into a *class-incremental* scenario (denoted as iCIFAR-100 by Rebuffi *et al.*, 2017b) by splitting the 100 classes into groups. In this study, we consider groups of 10 classes, thus obtaining 10 incremental batches.

The CNN model used for this experiment is the same as that used by Zenke *et al.* (2017) for experiments on the CIFAR-10/100 split (Maltoni and Lomonaco, 2019). It consists of four convolutional + two fully connected layers; details are available in Appendix A of Zenke *et al.* (2017). The model was pretrained on CIFAR-10 (Krizhevsky, 2009). Figure 6.3 compares the accuracy of the different approaches on iCIFAR-100. The results suggest that:

- unlike the naïve approach, *learning without forgetting* (LWF) (Li and Hoiem, 2016) and EWC provide some robustness against forgetting, even if in this incremental scenario their performance is not satisfactory. SI, when used in isolation, is quite unstable and performs worse than LWF and EWC.
- the accuracy improvement of CWR+ over CWR is very small here because the batches are balanced (so, weight normalization is not required), and the CNN initialization for the last-level weights was

already very close to 0 (we used the authors' default setting of a Gaussian with std = 0.005).
- ARR ($\lambda = 4.0e5$) consistently outperforms all the other approaches.

It is worth noting that both the experiments reported in Fig. 6.3 (i.e., an expanding (left) and fixed (right) test set, from left to right) lead to the same conclusions in terms of relative ranking among approaches; however, we believe that a fixed test set allows us to better appreciate the incremental learning trend and its peculiarities (saturation, forgetting, etc.) because the classification complexity (which is proportional to the number of classes) remains constant across the batches. For example, in the right graph, it can be noted that SI, EWC and LWF learning capacities tend to saturate after 6–7 batches, while CWR, CWR+ and ARR continue to grow; the same information is not evident on the left because

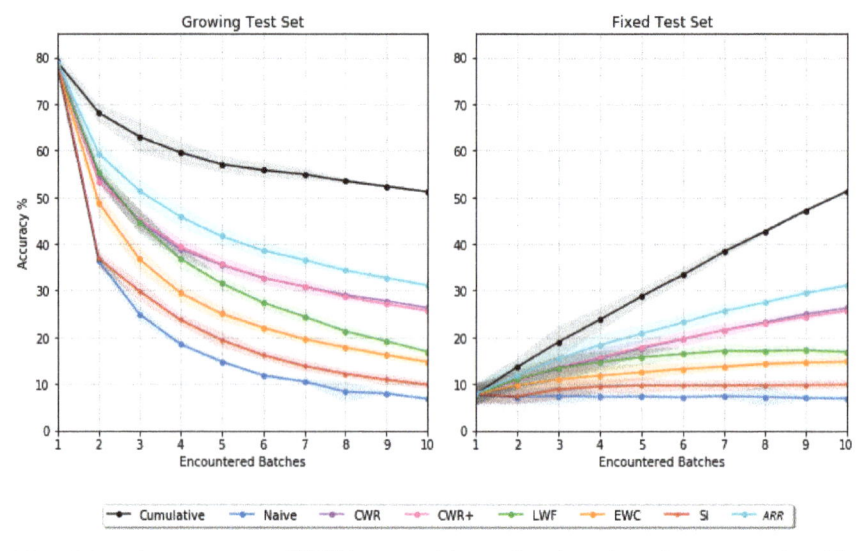

Fig. 6.3: Accuracy on iCIFAR-100 with 10 batches (10 classes per batch). Results are averaged over 10 runs: For all the strategies, hyperparameters have been tuned on run 1 and kept fixed in the other runs. The experiment on the right, consistent with the CORe50 test protocol, considers a fixed test set, including all 100 classes, while on the left we include in the test set only the classes encountered so far (analogously to the results reported by Rebuffi *et al.*, 2017b). Colored areas represent the standard deviation of each curve. Better viewed in color (Maltoni and Lomonaco, 2019).

of the underlying negative trend due to the increasing problem complexity.

Finally, note that absolute accuracy on iCIFAR-100 cannot be directly compared with that reported by Rebuffi et al. (2017b) because the CNN model used by Rebuffi et al. (2017b) is a ResNet-32, which is much more accurate than the model used here: On the full training set, the model used here achieves an accuracy of about 51%, while ResNet-32 achieves about 68.1%.

6.5.2 CORe50

While the accuracy improvement of the proposed approach with respect to the state-of-the-art rehearsal-free techniques have already been discussed in the previous section, a further comparison with other state-of-the-art continual learning techniques on CORe50 may be beneficial for better appreciating its practical impact and advantages in real-world continual learning scenarios and longer sequences of experiences. In particular, while ARR and ARR(α = *pool6*) have already been proven to be substantially better than LWF and EWC on the NICv2-391 benchmark (Lomonaco et al., 2020), a comparison with iCaRL (Rebuffi et al., 2017b), one of the best-known rehearsal-based techniques, is worth considering.

Unfortunately, iCaRL was conceived for *class-incremental* scenarios, and its porting to *class-incremental with repetition* (whose batches also include examples of known classes) is not trivial. To avoid subjective modifications, Lomonaco et al. (2020) began with the code shared by its original authors and emulated a *class-incremental with repetition* setting by: (i) always creating new virtual classes from examples in the coming batches; (ii) fusing virtual classes together when evaluating accuracies. For example, let us suppose we encounter 300 examples of class 5 in batch 2 and another 300 examples of the same class in batch 7. while two virtual classes are created by iCaRL during training, when evaluating accuracy, both classes point to the real class 5. Such an iCaRL implementation, with an external memory of 8,000 examples (much more than the 1500 used by the proposed latent replay but in line with the settings proposed in the original paper (Rebuffi et al., 2017b)), was run on NICv2-391; however, we were not able to obtain satisfactory results. In Fig. 6.4, we report the iCaRL accuracy over time and compare it

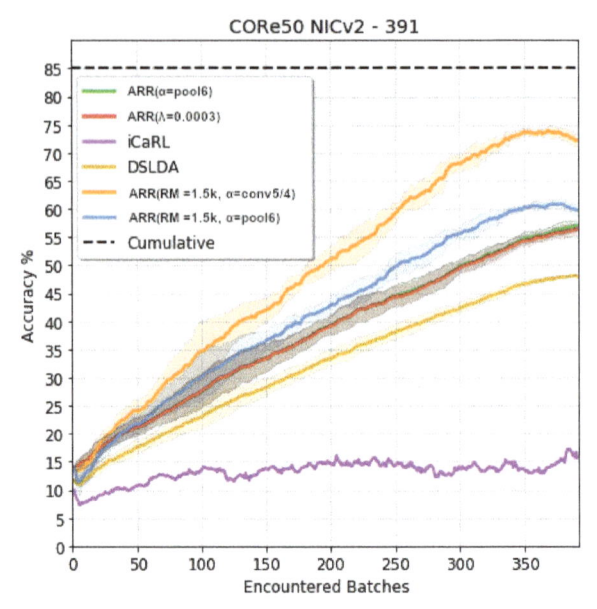

Fig. 6.4: Accuracy results on the CORe50 NICv2-391 benchmark of ARR($\alpha = pool6$), ARR($\lambda = 0.0003$), DSLDA, iCaRL, ARR($RM_{size} = 1500, \alpha = conv5\_4$) and ARR($RM_{size} = 1500, \alpha = pool6$). Results are averaged across 10 runs in which the batch order is randomly shuffled. Colored areas indicate the standard deviation of each curve. As an exception, iCaRL was trained only on a single run given its extensive run time ($\sim$14 days).

with ARR($RM_{size} = 1500, \alpha = conv5\_4/dw$), ARR($RM_{size} = 1500, \alpha = pool6$) as well as the top three performing rehearsal-free strategies introduced before: ARR($\alpha = pool6$), ARR($\lambda = 0.0003$) and DSLDA. While iCaRL exhibits better performance than LWF and EWC (as reported by Lomonaco et al., 2019b), it is far from DSLDA, ARR($\alpha = pool6$) and ARR($\lambda = 0.0003$).

Furthermore, when the algorithm has to deal with such a large number of classes (including virtual ones) and training batches, its efficiency becomes very low (as also reported by Maltoni and Lomonaco, 2019). In Table 1 of Lomonaco et al. (2020), the total run time (training and testing), memory overhead and accuracy difference with respect to the cumulative upper bound are reported. We believe that ARR($RM_{size} = 1500, \alpha = conv5\_4/dw$) represents a good trade-off in terms of efficiency–efficacy with a limited computational-memory overhead and only a $\sim$13% accuracy gap from the cumulative upper bound. For iCaRL, the total training time was ~14 days, compared to a training time of less than ~1 hour for the other learning algorithms on a single GPU.

6.5.3 ImageNet-1000

In order to further validate the ARR algorithm's scalability, Graffieti *et al.* (2022) performed a test on a competitive benchmark, such as ImageNet-1000, following the *class-incremental* benchmark proposed by Masana *et al.* (2020), which is composed of 25 experiences, with each of them containing 40 classes. The benchmark is particularly challenging due to the large number of classes (1000), the incremental nature of the task (with 25 experiences), and the data dimensionality of 224×224 (as with the ImageNet protocol).

In this experiment, Graffieti *et al.* (2022) tested ARR against both regularization-based methods (Dhar *et al.*, 2019; Kirkpatrick *et al.*, 2017; Li and Hoiem, 2016) and replay-based approaches (Belouadah and Popescu, 2019; Castro *et al.*, 2018; Chaudhry *et al.*, 2018a; Hou *et al.*, 2019; Rebuffi *et al.*, 2017b; Wu *et al.*, 2019). They used the same classifier (ResNet-18) and the same memory size for all the tested methods (20,000 examples, 20 per class); for the regularization-based approaches, the replay is added as an additional mechanism.

For ARR, they trained the model with an SGD optimizer. For the first experience, the algorithm was tuned with an aggressive learning rate of 0.1, a momentum of 0.9 and a weight decay of 10^{-4}. Then, the initial learning rate was multiplied by 0.1 every 15 epochs. The model was trained for a total of 45 epochs, using a batch size of 128. For all the subsequent experiences, SGD with a learning rate of 5×10^{-3} for the feature extractor's parameters ϕ and 5×10^{-2} for the classifier's parameters ψ were used. The model was trained for 32 epochs for each experience, employing a learning rate scheduler that decreases the learning rate as the number of experiences progresses. This was done to protect old knowledge against new knowledge when the former is more abundant than the latter. As in the first experience, the batch size was set to 128, composed of 92 examples from the current experience and 36 randomly sampled (without replacement) from the replay memory.

The results are shown in Table 6.1. Replay-based methods exhibit the best performance, with iCaRL and BiC exceeding a final accuracy

Table 6.1: Final accuracy on ImageNet-1000 following the benchmark of Masana et al. (2020) with 25 experiences composed of 40 classes each. For each method, a replay memory of 20,000 examples is used (20 per class at the end of training). Results for other methods reported by Masana et al. (2020).

Method	Final accuracy
Fine tuning (naive)	27.4
EWC-E (Kirkpatrick et al., 2017)	28.4
RWalk (Chaudhry et al., 2018a)	24.9
LwM (Dhar et al., 2019)	17.7
LwF (Li et al., 2019c)	19.8
iCaRL (Rebuffi et al., 2017b)	30.2
EEIL (Castro et al., 2018)	25.1
LUCIR (Hou et al., 2019)	20.1
IL2M (Belouadah and Popescu, 2019)	29.7
BiC (Wu et al., 2019)	32.4
ARR (Maltoni and Lomonaco, 2019)	**33.1**

of 30%. ARR($RM_{\texttt{size}} = 1500, \alpha = pool6$) outperforms all the baselines (33.1%), achieving state-of-the-art performance on this challenging benchmark and proving the advantage of flexible hybrid continual learning approaches.

However, considering that the top-1 ImageNet accuracy for ResNet-18 when trained on the entire dataset is 69.76%,[3] even for the best methods, the accuracy gap in the continual learning setup is very large. This suggests that continual learning, especially in complex scenarios with a large number of classes and high-dimensional data, is far from being solved, and further research should be devoted to this field.

6.5.4 *Replay layer selection*

In Fig. 6.5, we report the accuracy of ARR($RM_{\texttt{size}} = 1500, \alpha = \cdots$) for different choices of the rehearsal layer α for the CORe50 experiment.

---

[3] Accuracy taken from the torchvision official page: https://pytorch.org/vision/stable/models.html.

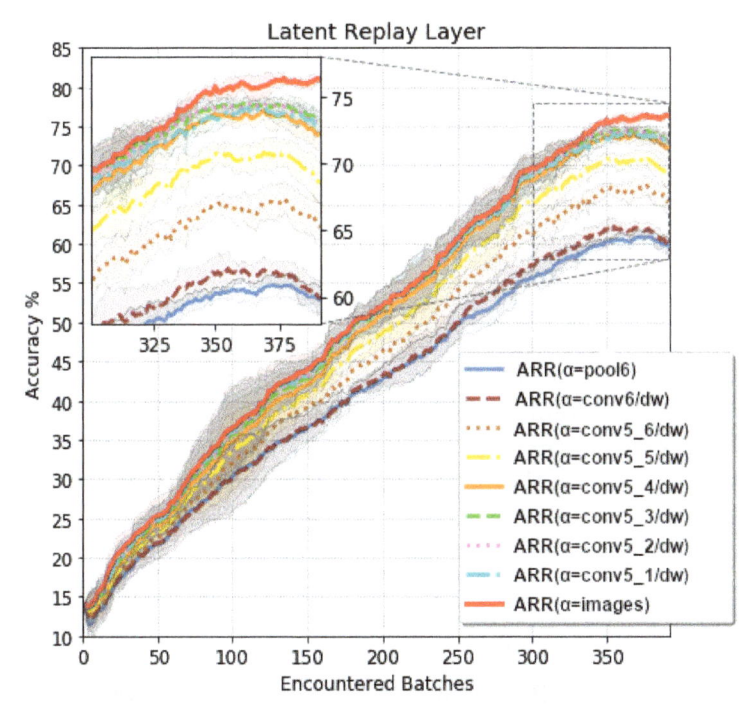

Fig. 6.5: ARR with latent replay ($RM_{size} = 1500$) for different choices of the latent replay layer. Setting the replay layer at the "images" layer corresponds to native rehearsal. The saturation effect which characterizes the last training batches is due to the data distribution in NICv2-391 (see Lomonaco *et al.*, 2019b): in particular, the lack of new instances for some classes (that already introduced all their data) slows-down the accuracy trend and intensifies the effect of activations aging.

As expected, when the replay layer is pushed down, the corresponding accuracy increases, proving that continual tuning of the representation layers is important. However, after `conv5_4/dw`, there is a sort of saturation, and the model accuracy is no longer improving. The residual gap (∼4%) with respect to native rehearsal is not due to the weight freezing of the lower part of the network but to the aging effect introduced above. This can be simply proved by implementing an "intermediate" approach that always feeds the replay pattern from the input and stops the backward at `conv5_4`: Such an intermediate approach achieved an accuracy at the end of the training very close to the native rehearsal (from raw data). We believe that the accuracy drop due to the aging effect can be further reduced with better tuning of BNR hyperparameters and/or with the introduction of a scheduling policy, making the global moment mobile windows wider as the continual learning progresses (i.e., more plasticity in the early stages and more stability later); however, such fine optimization is application specific and beyond the scope of this study.

To better evaluate the latent replay with respect to the native rehearsal, we report in Table 6.2 the relevant dimensions: (i) computation refers to the percentage cost in terms of operations of a partial forward (from the latent replay layer on) relative to a full forward step from the input layer; (ii) pattern size is the dimensionality of the pattern to be stored in the external memory (considering that we are using a MobileNetV1 with $128 \times 128 \times 3$ inputs to match the CORe50 image size); (iii) accuracy and Δ accuracy quantify the absolute accuracy at the end of the training and the gap with respect to a native rehearsal, respectively.

For example, `conv5_4/dw` exhibits an interesting trade-off because the computation is about 32% of the native rehearsal one, the storage is reduced to 66% (more on this point in Section 6.5.5), and the accuracy drop is mild (5.07%). ARR($RM_\text{size} = 1500, \alpha = pool6$) has a really negligible computational cost (0.027%) with respect to native rehearsal and still provides and an accuracy improvement of ~4% with respect to the non-rehearsal case (~60% vs. ~56%, as can be seen from Figs. 6.5 and 6.6, respectively).

Table 6.2: Computation–storage–accuracy trade-off with latent replay at different layers of a MobileNetV1 ConvNet trained continually on NICv2-391 with $RM_\text{size} = 1500$.

Layer	Computation (%) vs. native rehearsal	Example size	Final accuracy (%)	Δ Accuracy (%) vs. native rehearsal
Images	100.00	49152	77.30	0.00
conv5_1/dw	59.261	32768	72.82	−4.49
conv5_2/dw	50.101	32768	73.21	−4.10
conv5_3/dw	40.941	32768	73.22	−4.09
conv5_4/dw	**31.781**	**32768**	**72.24**	**−5.07**
conv5_5/dw	22.621	32768	68.59	−8.71
conv5_6/dw	13.592	8192	65.24	−12.06
conv6/dw	9.012	16384	59.89	−17.42
pool6	0.027	1024	59.76	−17.55

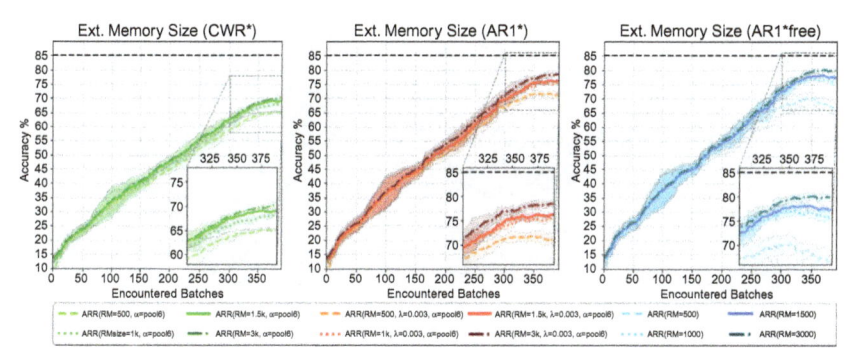

Fig. 6.6: Comparison of main ARR configurations on CORe50 NICv2-391 with different external memory sizes ($RM_\text{size} = 500, 1000, 1500$ and 3000 examples).

6.5.5 Replay memory size selection

To understand the influence of the external memory size, we repeated the experiment with different RM_size values: 500, 1000, 1500 and 3000. The results are shown in Fig. 6.6: It is worth noting that increasing the rehearsal memory leads to better accuracy for all the algorithms; however, the gap between 1500 and 3000 is not large, and we believe that 1500 is a good trade-off for this dataset. ARR($RM_\text{size} = ...$) works slightly better than ARR($RM_\text{size} = ...$, $\lambda = 0.003$) when a sufficient number of rehearsal examples are provided; however, as expected, accuracy is worse with light (i.e., $RM_\text{size} = 500$) or no rehearsal.

It is worth noting that the best ARR configuration in Fig. 6.6, i.e., ARR($RM_\text{size} = 3000$), is only 5% worse than the cumulative upper bound, and a better parametrization and exploitation of the rehearsal memory could further reduce this gap.

6.6 ARR Implementation in Avalanche

The ARR method we proposed in this chapter is the result of comprehensive reformalization of different variants and improvements proposed over the past few years, starting with Lomonaco and Maltoni (2017) and Maltoni and Lomonaco (2019). Original implementations of such methods (CWR, CWR+, CWR*, AR1, AR1* and AR1* with *latent replay*) exist in Caffe and PyTorch. However, given their diversity, it is quite difficult to move from one implementation to the

```
strategy = ARR(model, optimizer, criterion, mem_size, lambd, alpha)

for experience in benchmark.train_stream:

    strategy.train(experience)
    strategy.eval(benchmark.test_stream)
```

Fig. 6.7: `ARR` implementation in Avalanche. Given a set of hyperparameters, `ARR` can be instantiated and properly configured to be tested on a large set of benchmarks already available in Avalanche.

next and apply them to settings and scenarios even slightly different from the one on which they have been proposed.

In order to exploit the general applicability and flexibility of the `ARR` method, we decided to reimplement it directly in Avalanche (Lomonaco et al., 2021). *Avalanche*, an open-source (MIT licensed) end-to-end library for continual learning based on PyTorch, devised to provide a shared and collaborative codebase for fast prototyping, training and evaluation of continual learning algorithms.

Thanks to the Avalanche portable implementation (soon to be integrated in the next stable version of the library), `ARR` can be configured to reproduce the experiments presented in this chapter (Fig. 6.7), conforming to the previously proposed strategies (e.g., `AR1*` and `CWR*`), as well as being ready to be tested on a large set of benchmarks already available in `Avalanche` or that can be easily added to the library.

6.7 Conclusion

In this chapter, we showed that `ARR` is a flexible, effective and efficient technique to continually learn new classes and new instances of known classes, even from small and non-i.i.d. batches. `ARR`, instantiated with latent replay, is indeed able to learn efficiently, and at the same time, the achieved accuracy is not far from the cumulative upper bound (about 5% in some cases). The computation–storage–accuracy trade-off can be defined according to both the target application and the available resources so that even edge devices with no GPUs can learn continually. Moreover, ARR can easily be extended to support more sophisticated replay memory management strategies (also to contrast the *aging effect*) and even be coupled with a generative model trained in the loop and capable of providing pseudo-activation volumes on demand, as initially shown by Graffieti *et al.* (2022).

© 2024 World Scientific Publishing Company
https://doi.org/10.1142/9789811286711_0007

Chapter 7

Task-Agnostic Inference Using Base–Child Classifiers

Pranshu Ranjan Singh[*,†], Saisubramaniam Gopalakrishnan[*,‡],
Savitha Ramasamy[*,§] and ArulMurugan Ambikapathi[*,∥]

[*]*Institute for Infocomm Research, Agency for Science,
Technology and Research (A*STAR), Singapore*
[†]*pranshu96ranjan@gmail.com*
[‡]*saisubramaniam147@gmail.com*
[§]*ramasamysa@i2r.a-star.edu.sg*
[∥]*a.arulmurugan@gmail.com*

Abstract

Continual learning (CL) aims to learn new tasks through the forward transfer of information learned from previous tasks without forgetting them. In task-incremental CL, task information is vital during both strategy development and inference. Providing such partial knowledge about the test sample demands additional complexity and may become intractable, especially when the sample source is ambiguous. In this work, we design a task-agnostic inference approach that uses a base–child hybrid setup to incrementally learn tasks while mitigating forgetting. Multiple base classifiers guided by reference points learn new tasks, and this information is distilled via a feature space-induced sampling strategy. A central child classifier consolidates information across tasks and infers the task identifier automatically. Experimental results on standard datasets show that the proposed approach outperforms the various state-of-the-art regularization and replay CL algorithms in terms of accuracy by 50% and 7% with homogeneous and heterogeneous tasks, respectively, in task-agnostic scenarios.

7.1 Introduction

Deep learning (DL) methods have shown state-of-the-art results for various supervised and unsupervised tasks and on complex datasets, such as ImageNet (Deng et al., 2009) and COCO (Lin et al., 2014). A lot of research had been focused on improving deep neural network architectures, starting with AlexNet (Krizhevsky et al., 2012), which paved the way forward in this direction. Subsequently, deep architectures, such as Inception (Szegedy et al., 2015), ResNet (He et al., 2016) and DenseNet (Huang et al., 2017), further pushed the performance benchmark on image classification. Although, DL methods have revolutionized many tasks in different domains, such tasks are mostly *specific* or *standalone*. Most of the success of DL methods has been targeted toward such *specific* tasks as classification, regression or segmentation. There has been limited research on learning a sequence of tasks. However, recent advances in this direction have garnered the attention of the research community (Rusu et al., 2016b; Kirkpatrick et al., 2017; Lopez-Paz and Ranzato, 2017).

When a DL model trained on a task is then trained on another task, it *forgets* how to perform the first task and demonstrates poor results on the first task. This phenomenon is known as *catastrophic forgetting*, and it imposes a serious concern when learning a sequence of tasks (Goodfellow et al., 2013). A variant of the above scenario, where the first task is a *source* (trained on a generic dataset) and the second task is a *target* (trained on a specific dataset), finds itself a use case in transfer learning (TL) (Zhuang et al., 2020). In TL, the goal is to *fine-tune* the previously learned DL model of the source to a target task. As good performance on the first task (*source*) is not required for TL, catastrophic forgetting does not hinder the learning.

Continual learning (CL) allows neural networks to learn a sequence of tasks without catastrophic forgetting of preceding tasks and learn new tasks through the forward transfer of information. The concepts of knowledge retention (minimizing the forgetting of previous tasks when learning future tasks), knowledge transfer (reusing knowledge acquired from previous tasks to solve current and future tasks efficiently and enable learning) and parameter efficiency (the number of parameters in the model is bounded or grows at most sublinearly with newer tasks) are some of the other desired properties for a CL model (Sodhani et al., 2020).

Forgetting in CL may arise under different scenarios: an increase in the number of tasks (*task-incremental*), a shift in domain or data characteristics (*domain-incremental*), or an increase in the number of classes (*class-incremental*) (Van de Ven and Tolias, 2019; Kiyasseh et al., 2020). In a task-incremental setting, a sequence of tasks (each comprising multiple classes) is trained, and task identity is known during test time. Typically, a multi-head output layer DL model is used, and it trains each head corresponding to a specific task. In a domain-incremental setting, the structure of tasks remains the same while the input distribution keeps changing. A single-head output layer DL model is used, and test identity is not required during inference. In a class-incremental setting, a sequence of classes (tasks) is learned; however, task identity is not provided during inference. Generally, methods adopt a single-head output layer DL model, with each incoming class being added to the output layer (Van de Ven and Tolias, 2019).

Regardless of the type of CL scenario, CL of sequential tasks is predominantly accomplished through architectural (Rusu et al., 2016b; Yoon et al., 2018), regularization (Kirkpatrick et al., 2017; Zenke et al., 2017; Nguyen et al., 2018; Ebrahimi et al., 2020a) or replay/rehearsal (Isele and Cosgun, 2018; Lopez-Paz and Ranzato, 2017; Tao et al., 2020; Rao et al., 2019; van de Ven et al., 2020) strategies. There are methods which use a combination of these strategies as well (Maltoni and Lomonaco, 2019; Li and Hoiem, 2018; Singh et al., 2021).

Architectural strategies primarily focus on optimized network utilization and allow for growing and pruning of neurons or layers in the DL model (Rusu et al., 2016b; Yoon et al., 2018; Ye and Bors, 2021; Yang et al., 2022). Regularization strategies use a fixed DL model and add an additional regularization loss for remembering previous tasks, thereby finding parameters for the model that work well for all seen tasks (Kirkpatrick et al., 2017; Zenke et al., 2017; Nguyen et al., 2018; Ebrahimi et al., 2020a). Regularization is usually achieved through penalizing the large deviations in the network parameters that are trained for preceding tasks while learning successive tasks. Regularization is also achieved through replay in some cases, such as in embedding regularization (Pomponi et al., 2020), where it is achieved through the replay of internal feature embedding in the stored network. Similarly, LwF (Li and Hoiem, 2018) exploits knowledge distillation and fine-tuning to replay class predictions (as soft labels) of past tasks obtained from current task samples. Rehearsal or replay strategies preserve the knowledge of previous tasks by either retaining original coreset samples or learning a generative model over previous tasks and replaying those samples (either the original coreset or via the generative model) when training for future tasks (Lopez-Paz and Ranzato, 2017; Tao et al., 2020; Isele and Cosgun, 2018; Shin et al., 2017; Egorov et al., 2021). Similarly, the magnitude and direction of gradients are penalized for strong adaptations in GEM (Lopez-Paz and Ranzato, 2017) and A-GEM (Chaudhry et al., 2018b).

The CL scenarios discussed above have a clear separation of task boundaries when presented for training and may use task identity/information when performing inference (during test time) (Van de Ven and Tolias, 2019). For a task-incremental setting, the developed methods use task identity to perform inference at the test stage. In this chapter, the method that is discussed performs task-agnostic inference for task-incremental problems and automatically infers the task identity/identifier (ID). Although there are methods that consider the absence of task ID during the training stage as well (He et al., 2020; Zeno et al., 2018), the scope of this chapter is limited to those that use task ID during the training stage.

This chapter is an extension to the work "Task-Agnostic Continual Learning Using Base-Child Classifiers" (Singh et al., 2021). It presents a task-agnostic inference method for performing CL on a sequence of classification tasks (task-incremental) and lies at the intersection of regularization and replay strategies. The method uses *a dual network architecture*. It must be noted that dual networks have been used to modularize learning components, where new information is actively collected by one network and consolidated by a central and stable secondary network (Tarvainen and Valpola, 2017; Mnih et al., 2013). This method comprises a base classifier and a child classifier; hence, the method is named *base–child classifiers (BCCs)*. The base classifier is trained for each incoming task in the sequence whose instances are created for each task. A central child classifier, which is continually trained, consolidates the knowledge from multiple tasks.

Once the training of the base classifier is completed for a given task, the knowledge of the individual task is *distilled* through sampling boundary point pairs (image, latent space) of each class from the latent space (LS) of the base classifier. The child classifier is then continually trained by rehearsing/replaying the saved boundary point samples for each class. Thus, knowledge distillation, replay through class boundary points sampling and regularization are exploited for continual learning of a sequence of classification tasks through a dual network architecture. The key contributions of this work include:

- effective coexistence and retention of knowledge, enabling intra- and inter-task separation using reference points;
- dual network setup for CL and training methodology, modular base and child classifier networks for shared representations;
- boundary points sampling techniques in high-dimensional LS for selective replay;
- automatic task identification during inference (task-agnostic inference) using reference points;
- experimental demonstration on standard datasets that BCCs outperform the various state-of-the-art regularization and replay CL algorithms in task-agnostic inference scenarios.

The rest of the chapter covers the main methodology (Section 7.2) and experimental results and discussion (Section 7.3) in detail.

7.2 Task-Incremental Learning Using Base–Child Classifiers

This section covers the formulation for BCC and describes each of its components in depth. Section 7.2.1 presents the mathematical formulation for the CL task-incremental classification problem. Section 7.2.2 provides an overview of the method through a high-level block diagram. The ensuing sections cover each step of the method in detail. Section 7.2.3 describes the importance of reference points and the criteria for obtaining them. Section 7.2.4 covers the loss formulation for optimizing base classifiers. Sections 7.2.5 and 7.2.6 explain the use of boundary points sampling and various sampling techniques to obtain them, respectively. LS reconstruction and its utilization in constructing child classifiers are presented in Sections 7.2.7 and 7.2.8, respectively. This is followed by Section 7.2.9, which describes the automated task inference strategy that enables task-agnostic inference for BCC. Finally, Section 7.2.10 combines the various blocks of BCC and presents the algorithm to perform continual learning using base–child classifiers.

7.2.1 Problem formulation: Task-incremental classification

Consider the CL context with a sequence of N classification tasks $\{T_1, T_2, \ldots, T_k, \ldots, T_N\}$ from the same or different datasets. When the given sequence of tasks is chosen from the same dataset, each task corresponds to a few (at least two) class classifications. Each task T_k comprises N_k training samples and N_{T_k} class labels. Let $D_{T_k} = \{(\mathbf{X}_1, y_1), (\mathbf{X}_2, y_2), \ldots, (\mathbf{X}_{N_k}, y_{N_k})\}$ be the set of training samples (collection of (image, label) tuples, each containing input image and the corresponding class label) for task T_k, with input image samples $\mathbf{X}_i \in \Re^{a \times b \times c}$ (a rows, b columns, and c channels for the image sample) having class labels $\forall y_i \in \{0, 1, \ldots, N_{T_k} - 1\}$.

The DL model is trained in a CL fashion, i.e., when the DL model M_{t-1} (obtained at the end of training task T_{k-1}) is optimized for classification task T_k to obtain model M_t, it is trained in such a way that the previous tasks' knowledge ($T_1, T_2, \ldots, T_{k-1}$) is useful while learning the current task T_k, and most importantly, it does so with minimal forgetting of the previous tasks. Since each task T_k differs with the nature of data, the number of samples N_k, and class labels y, a naive approach where the new successive task training overwrites existing network weights will not be able to achieve the set goals of continual learning, and hence, a CL strategy is required to learn such a sequence of tasks.

7.2.2 Base–child classifiers

The BCC approach overcomes the challenge of continual learning described in Section 7.2.1 by dividing the workload of continual task-incremental classification into multiple task-specific base classifiers and a central consolidated child classifier, as illustrated in Fig. 7.1. A set of reference points, corresponding to each class present in each

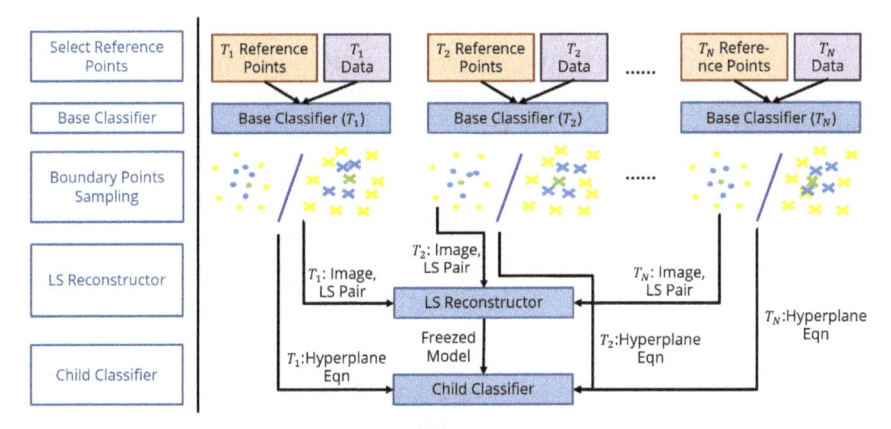

Fig. 7.1: Block diagram for BCCs. This depicts the various components of BCC and describes the method for performing task-incremental classification.

task, is selected under specified constraints and is used later for training base classifiers and automated task inference. Base classifiers are trained using a weighted loss combination of cross-entropy and clustering loss (Ghosh and Davis, 2018). The clustering loss is defined as the mean absolute error (MAE) between the LS representations and previously selected class-specific reference points. The above strategy ensures that the classification is performed in such a manner that classes within the current task are well separated (intra-task, inter-class separation) and individual base classifiers of different tasks are also well separated (inter-task separation).

After training the base classifier for the current task, BCC employs a boundary points sampling technique (farthest from each LS representation of the base classifier), specifically designed to extract useful knowledge in the form of *(image data sample, latent representation)* tuples from the trained base classifier. With such a collection of sampled tuples from various base classifiers (for all tasks observed until now), an LS reconstructor model is continually trained to learn the shared representations corresponding to all observed tasks. Appending the frozen LS reconstructor model with the hyperplane separation equations (classification head or output layer) of each task (from the corresponding base classifier), the central child classifier is constructed. There is no training requirement for the child classifier, as it builds upon two sets of pretrained model parameters, the LS reconstructor model and the classification head of base classifiers, and hence can be directly used for inference. The utilization of selected reference points in the base classifier training encourages strong separation among different tasks (inter-task separation). This knowledge can also be used to perform task inference by comparing the distance between LS representations of test image sample and reference points. Thus, reference points in the BCC setup enable the coexistence of knowledge and automatic task inference (task-agnostic) during the testing phase. Section 7.3 demonstrates the efficacy of BCC methodology as compared to other baseline methods on homogeneous and heterogeneous datasets.

7.2.3 Selection of reference points

Reference points are created/obtained for each class in a given task to ensure well-defined inter-class separation as well as inter-task separation. These are fixed points (in the LS dimension) that act as prespecified class means/centroids in the LS for all classes. They are used during training of the base classifier by enforcing that the classification LS clusters of each class are centered around their corresponding reference points. For the given task T_k with N_{T_k} classes (having class labels 0, 1,..., $N_{T_k} - 1$), let the LS dimension of the base classifier be s, i.e., *LS representations* $\in \Re^s$. For each class in task T_k, a reference point is chosen (handcrafted feature in LS), forming the table R_{T_k} of dimension $N_{T_k} \times s$. For task T_k, the reference points should ensure that they are well separated from the reference points of the previous tasks ($R_{T_1}, \ldots, R_{T_{k-1}}$). Figure 7.2(a) shows the t-SNE visualization of initial reference points selected for 10 classes of the MNIST dataset. Figure 7.2(b) shows the t-SNE visualization of the classification LS obtained from the base classifier trained on the MNIST dataset using those initial reference points (red dots/points in the figure).

For a two-task scenario, one can choose all positive LS representations of the form $(z_1, z_2, \ldots, z_s) \in \Re^s$ such that $\forall z_i \geq 0$ to represent reference points for one task and all negative LS representations with $\forall z_i \leq 0$ to represent the reference points of the other task. The above scheme ensures that tasks are well separated. To enable inter-class separation within each task, any point having, say, a *unit* distance from the origin can be selected. One straightforward selection is to have one of the z_i components set to +1 or −1 and

Task-Agnostic Inference Using Base–Child Classifiers 131

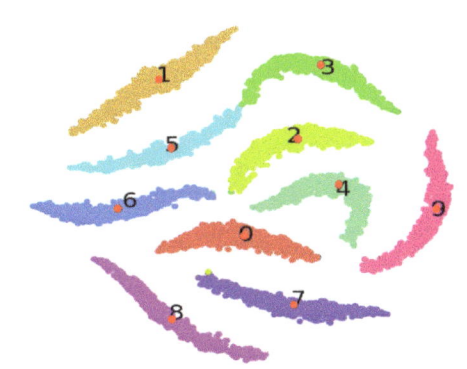

(a) *t*-SNE visualization of 10 chosen reference points corresponding to each of the 10 classes of the MNIST dataset.

(b) *t*-SNE visualization of classification LS for 10 MNIST classes and selected reference points (red dots).

Fig. 7.2: *t*-SNE visualization for reference points R_{T_k} and its effect on the trained base classifier's LS of the MNIST dataset.

all others set to 0. Theoretically, such a selection mechanism allows for a very large number of classes (at least s using the straightforward selection scheme) to be represented via reference points. For a more general setting, where multiple tasks (number of tasks >2) are present, one can choose reference points for each task to lie in one of the 2^s regions created by s axes in $\Re^s$-dimension space. To get the reference points for each class of a task, a set of equidistant points in one of the 2^s regions can be selected. This strategy ensures inter-task separation, as all the 2^s regions are well separated and inter-class separation through the construction of equidistant reference points within each region corresponding to tasks.

7.2.4 *Base classifier*

The base classifier is a supervised classifier network that performs classification exclusively for the current task T_k. Given the training samples D_{T_k} for task T_k, the base classifier is optimized using a weighted sum of two losses:

- Cross-entropy loss (applied on softmax/class probabilities)

$$L_{\text{CE}} = -\frac{1}{N_k}\sum_{i=1}^{N_k}\mathbf{y}_i.\log(\hat{\mathbf{y}}_i). \tag{7.1}$$

- Clustering loss (applied to latent space), the MAE between latent space representations and a class-specific reference point:

$$L_{\text{MAE}} = \frac{1}{N_k} \sum_{i=1}^{N_k} \left| \hat{\text{ls}}_i - R_{T_k}[y_i] \right|, \quad (7.2)$$

where $\hat{\mathbf{y}}_i$ is the predicted class probability vector of N_{T_k} dimensions, $\mathbf{y}_i$ is the ground truth one-hot encoded vector, y_i is the ground truth class label ($\mathbf{y}_i$ is the one-hot encoded vector for y_i), $\hat{\text{ls}}_i$ is the predicted LS vector from the base classifier and $R_{T_k}[y_i]$ specifies the class-specific reference point of class label y_i of task T_k, for the ith training sample. Then, the final weighted loss function is given by

$$L_{\text{weighted}} = w_1 . L_{\text{CE}} + w_2 . L_{\text{MAE}}, \quad (7.3)$$

where w_1 and w_2 are the respective weights for the two losses: cross-entropy and clustering. A block diagram for the base classifier is depicted in Fig. 7.3. Once the training of the base classifier for task T_k is completed, the classification head (separating hyperplane

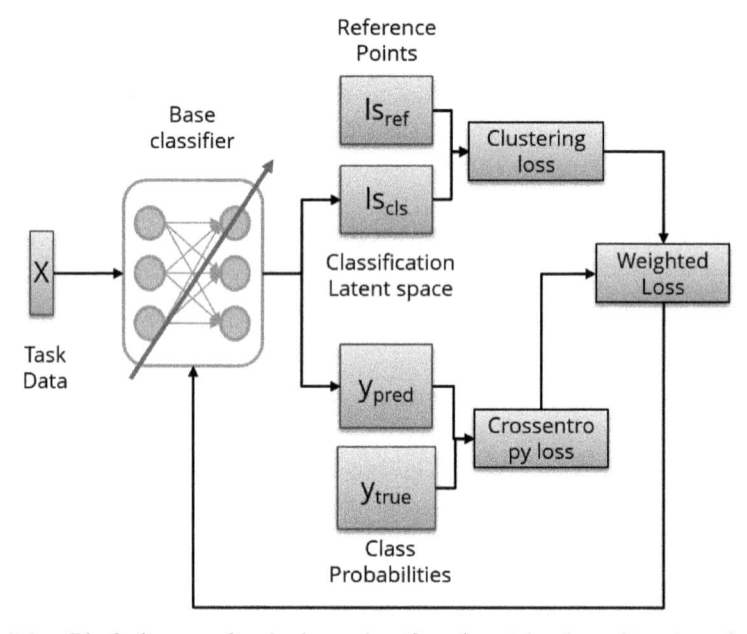

Fig. 7.3: Block diagram for the base classifier: A weighted combination of cross-entropy and clustering loss is used for optimizing the base classifier network.

equations or output layer weights) of task T_k is stored (later used in the child classifier).

7.2.5   Boundary points sampling

Upon training the base classifier for a given task T_k, selective representative image samples from D_{T_k} and their corresponding LS representations are stored in memory. These samples will be later used for replay/rehearsal while training the continual LS reconstructor model. BCCs use a *boundary points sampling strategy* to locate $(\mathbf{X_j}, \hat{\mathbf{ls}}_j)$ tuples situated at the boundary of each class cluster by selecting the top $p\%$ samples whose LS vectors are farthest from one another in the training set. The farthest from each point sampling method is described in Algorithm 7.1.

Using Algorithm 7.1, $S_{T_k} = \{(\mathbf{X}_j, \hat{\mathbf{ls}}_j)\}$ is obtained as a set of selected boundary point samples, where $\mathbf{X_j}$ is the image/data sample and $\hat{\mathbf{ls}}_j$ is the corresponding LS vector for task T_k. Figure 7.4 shows the t-SNE visualization for the classification LS of class label 0 (airplane) on the Cifar10 dataset. The *blue* cluster is obtained after training the base classifier on Cifar10, and the *red* points on the boundary of the cluster indicate the selected samples. Figure 7.5 illustrates the t-SNE visualization for all 10 Cifar10 classes as well as selected boundary points. The clusters numbered from 0 to 9 correspond to cluster centers for training samples of corresponding classes, and the clusters numbered from 11 to 19 indicate the cluster centers for the selected boundary points of corresponding classes.

7.2.6   Various sampling methods

Boundary points sampling methodology is an important component of BCCs, as it provides the basis for training the next component, the LS reconstructor, and thereby the child classifier. In Section 7.2.5, the farthest from each point sampling technique was discussed in order to obtain the boundary point samples. However, there were other sampling strategies that were also considered, and the setup of BCC allows for any sampling method to be used as long as the selected samples obtained enable a good child classifier model. This section describes other sampling methods that were explored as potential

Algorithm 7.1 Farthest from each point sampling method.

Require: $Q_{T_k} = \left\{(\mathbf{X}_i, \hat{\mathbf{ls}}_i, y_i)\right\}_{i=1}^{N_{T_k}}$: Set of (image, latent space, class label) tuples for task T_k

Require: p: Fraction of samples to be selected per class

1: $S_{T_k} \leftarrow \{\ \}$, where S_{T_k} denotes the set of selected boundary points samples
2: **for** $c \in \{0, 1, \ldots, N_{T_k} - 1\}$ **do**
3:    $Q_{T_k}^c \leftarrow \left\{(\mathbf{X}_j^c, \hat{\mathbf{ls}}_j^c, y_j): y_j = c\right\} \subset Q_{T_k}$
4:    $S_{T_k}^c \leftarrow \{\ \}$, where $S_{T_k}^c$ denotes the set of selected boundary points samples for class label c
5:    Let $\mathbf{ls}_{\text{mean}}$ be the mean latent space for class label c
6:    Compute $\mathbf{ls}_{\text{max}} \leftarrow \arg\max_{\hat{\mathbf{ls}}} \left\|\hat{\mathbf{ls}}_j - \mathbf{ls}_{\text{mean}}\right\|_2$
7:    $S_{T_k}^c \leftarrow S_{T_k}^c \cup \{(\mathbf{X}_{\text{max}}, \mathbf{ls}_{\text{max}})\}$, where $\mathbf{X}_{\text{max}}$ is the image sample corresponding to $\mathbf{ls}_{\text{max}}$
8:    $Q_{T_k}^c \leftarrow Q_{T_k}^c - \{(\mathbf{X}_{\text{max}}, \mathbf{ls}_{\text{max}}, c)\}$
9:    **for** $r \in \left\{1, \ldots, p.len(Q_{T_k}^c) - 1\right\}$ **do**
10:      Compute $\mathbf{ls}_{\text{max}} \leftarrow \arg\max_{\hat{\mathbf{ls}}} \left\|\hat{\mathbf{ls}}_j - \mathbf{ls}_{\text{max}}\right\|_2$
11:      $S_{T_k}^c \leftarrow S_{T_k}^c \cup \{(\mathbf{X}_{\text{max}}, \mathbf{ls}_{\text{max}})\}$, where $\mathbf{X}_{\text{max}}$ is the image sample corresponding to $\mathbf{ls}_{\text{max}}$
12:      $Q_{T_k}^c \leftarrow Q_{T_k}^c - \{(\mathbf{X}_{\text{max}}, \mathbf{ls}_{\text{max}}, c)\}$
13:    **end for**
14:    $S_{T_k} \leftarrow S_{T_k} \cup S_{T_k}^c$
15: **end for**
16: **return** S_{T_k}

sampling techniques for BCC. In Section 7.3.5, a comparison between different sampling approaches is provided, and the farthest from each point sampling technique outperforms other sampling methods.

Let $Q_{T_k} = \{(\mathbf{X}_i, \hat{\mathbf{ls}}_i, y_i)\}_{i=1}^{N_{T_k}}$ be the set of (image, latent space, class label) tuples corresponding to the training samples of task T_k. Samples will be drawn from this set using various sampling methods to obtain $S_{T_k} = \{(\mathbf{X}_j, \hat{\mathbf{ls}}_j)\}$ as the set of selected samples.

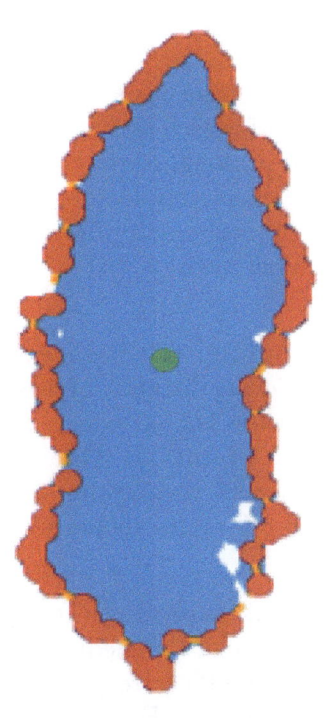

Fig. 7.4: Selected points (red) shown along with training samples (blue) and the class mean (green) using t-SNE visualization for class 0 (airplane) of the Cifar10 dataset.

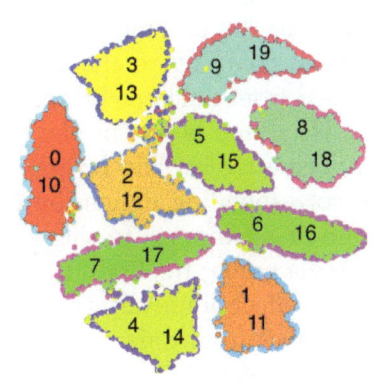

Fig. 7.5: Points lying on the boundaries (shown in different colors) of each class cluster are selected boundary points for the Cifar10 dataset. Labels 0–9 indicate the 10 classes in Cifar10, and labels 10–19 correspond to the selected boundary points of those classes.

Random sampling: Randomly choose $p\%$ samples uniformly from each class/overall, where p is the percentage of samples required. This selection can be done from each class or over all the classes together. When performed for each class individually, it ensures that S_{T_k} contains samples proportional to the number of samples present in each class. However, when samples are drawn from the overall mixture of classes, this proportionality between individual classes is not guaranteed to be maintained.

Farthest distance sampling: Select $p\%$ samples from each class with the maximum Euclidean distance between LS representations $\hat{\mathbf{ls}}_i$ and the mean LS vector $\mathbf{ls}_{\text{mean}}$, where p is the percentage of samples required. Let $d = \{d_j : d_j = \|\hat{\mathbf{ls}}_j - \mathbf{ls}_{\text{mean}}\|_2\}$ be the set containing Euclidean distances, sorting this set and selecting samples corresponding to the top $p\%$ values generates S_{T_k}.

Univariate combinations sampling: For each class, obtain the minimum, maximum and mean (or median) LS values for each LS dimension, and select the samples corresponding to all three measures (minimum, maximum and mean). For the minimum and maximum LS measures, there exits a unique sample lying in the training set; however, the mean LS value may not correspond to any available sample in the training set. In that case, the sample lying closest to the mean LS value (having the minimum distance with respect to the mean LS value) is selected. There can be overlap between the selected samples obtained from different LS dimensions.

Formally, let $\mathbf{ls}_{\min}$, $\mathbf{ls}_{\max}$ and $\mathbf{ls}_{\text{mean}}$ be the minimum, maximum and mean LS vectors, respectively. The LS representations $\hat{\mathbf{ls}}_i$ having LS values in the corresponding dimension close (exactly equal in the case of minimum or maximum) to these three LS vectors are selected and combined (via union) to obtain the set S_{T_k}. As the selection criteria prioritize the effect of univariate (each individual dimension) measures, and samples from different measures are combined to give the final set, the approach is termed *univariate combinations sampling*. A schematic diagram of the selection process is provided in Fig. 7.6. It shows a simple example of a 2-*dim* LS, where selected samples (orange blobs) correspond to minimum, maximum or mean measures per LS dimension.

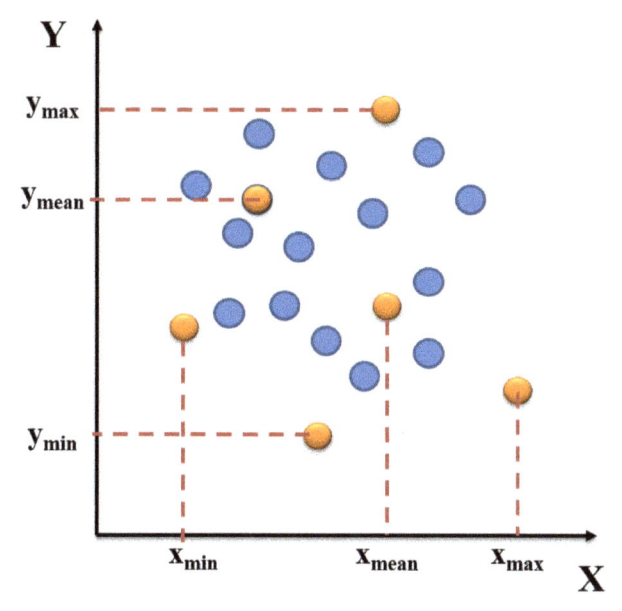

Fig. 7.6: Univariate combinations sampling method. Blue and orange blobs indicate the training data points and selected samples, respectively.

Disjoint univariate combinations sampling: As previously described, the univariate combinations sampling method selects samples over the entire LS of a class. This technique will pick good representative samples when the classification LS is dense and forms a single large cluster. However, there could be scenarios (e.g., high-dimensional LS for complex datasets) where the LS is sparse and there exist multiple clusters of points within a single class. Hence, disjoint univariate combinations sampling extends the previous method by first partitioning the LS of a class into r clusters (regions) and then executing univariate combinations sampling on each of the obtained clusters. Finally, the disjoint set of selected samples from r clusters is merged to obtain the set S_{T_k}. K-means or DBSCAN (Ester et al., 1996) can be used to perform the initial clustering on LS representations. A straightforward technique for obtaining clusters is to partition the LS points into concentric hyperspheres of different radii, thereby obtaining the smallest hypersphere (the least-radius hypersphere) and multiple hyperannuli (regions between concentric hyperspheres) as disjoint clusters.

Outermost cluster sampling: This method is similar to the previously described disjoint univariate combinations sampling. It also performs an initial clustering to restrict the sample space for selection. Perform concentric clustering by dividing the LS space into concentric hyperspheres and selecting the outermost hyperannuli (clusters). The set of selected samples S_{T_k} is obtained by performing random sampling to get the desired number of boundary point samples. Figure 7.7 depicts sample selection for a *2-dim* LS example using the outermost cluster sampling method. The concentric circles C_1, C_2 and C_3 obtained from respective radii r_1, r_2 and r_3, and centers (green dot) as the mean LS vector, partition the LS into three regions (ignoring the external region lying outside of C_3). As boundary points will lie in the outermost region, i.e., a set of points bounded by circles C_2 and C_3, they are chosen as the selected samples.

Low-density Gaussian distribution sampling: For each class of LS samples, fit a multi-variate Gaussian distribution over the classification LS and select the samples that fall in the

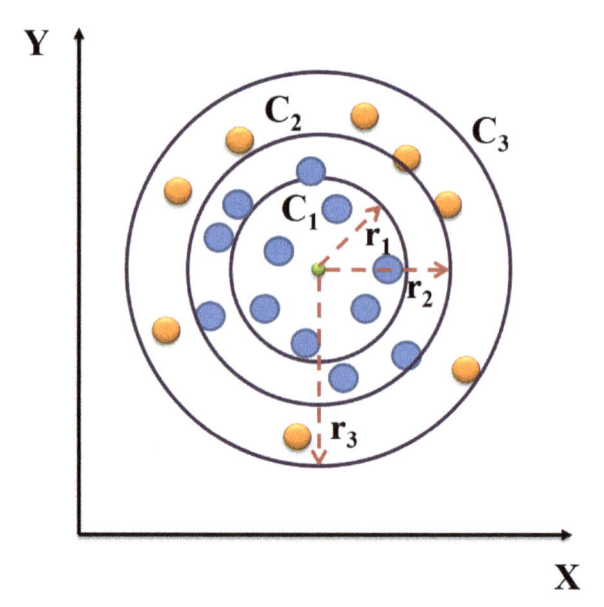

Fig. 7.7: Outermost cluster sampling method. Blue and orange blobs indicate the training data points and selected samples, respectively.

tail (*low-density* region). Compute the sample mean vector and covariance matrix from the training samples of each class label, thereby obtaining a multi-variate Gaussian distribution for each class label. Measure the density of class-specific training samples using the corresponding Gaussian probability density function and select the bottom $p\%$ samples having the lowest density values, where p is the percentage of samples required. This method is similar to the outermost cluster sampling approach, as both select samples that are extreme. However, this strategy differs from the former as it can choose points present in sparser regions (when there are multiple clusters), whereas the previous method will work well when there exists a single cluster per class (less sparsity in classification LS).

***t*-SNE projections based convex/concave hull sampling:** So far, all the discussed methods performed the sampling in the original high-dimensional latent space. This method performs sample selection by first projecting the original high-dimensional latent space to a low-dimensional (2-*dim*) *t*-SNE space (which is generally used for visualization of the LS) (Van der Maaten and Hinton, 2008) and then selecting samples corresponding to convex hull and concave hull (alpha shape) for each class label. A convex hull of q set of points (forming some shape) is the smallest convex set that contains all those given points, whereas a concave hull (Moreira and Santos, 2007) is the set of points that best describes the region occupied by a given set of points and hence captures the boundary of the region. However, unlike convex hulls, there are multiple possible configurations for concave hulls, which are controlled by some hyperparameters. Figure 7.8 illustrates the samples selected using the convex hull (left) and concave hull (right) strategies for class 0 (*airplane* class) of the Cifar10 dataset. It can be observed that the convex hull omits most of the boundary points, whereas the concave hull does a great job of identifying/selecting most of the boundary samples.

7.2.7 *LS reconstructor*

Boundary points samples, S_{T_1}, S_{T_2}, ..., S_{T_k} selected over the past task T_1, T_2, ..., T_k are accumulated to form a joint training set

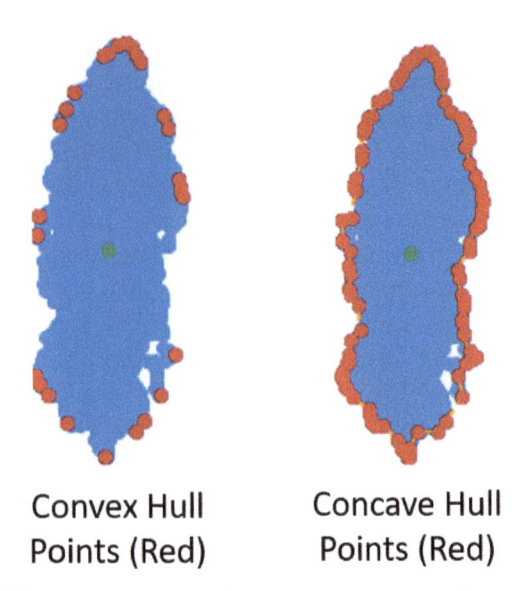

Fig. 7.8: *t*-SNE projections-based convex/concave hull sampling method. Selected points (red) using convex hull (*left*) and concave hull (*right*) from the *t*-SNE classification LS of the Cifar10 dataset class 0 (*airplane*).

and then distilled to an LS reconstructor model. Through knowledge distillation, the LS reconstructor model is able to consolidate the knowledge acquired by individual base classifiers into a central knowledge base. This enables the LS reconstructor model to learn shared representations in LS across all tasks.

Let S_{all} be defined as the collection of all the boundary points (across all observed tasks until T_k), i.e.,

$$S_{\text{all}} = \bigcup_{i=1}^{k} S_{T_i}. \tag{7.4}$$

The LS reconstructor is then *continually* trained as a regression network over S_{all} training samples to map an image sample $\mathbf{X_i}$ to its corresponding LS $\hat{\mathbf{ls}}_i$ by optimizing for MAE between the saved LS $\hat{\mathbf{ls}}_i$ and the predicted output $\mathbf{ls}_i'$. The loss function is described as follows:

$$L_{\text{rec}} = \frac{1}{n(S_{\text{all}})} \sum_{i=1}^{n(S_{\text{all}})} |\mathbf{ls}_i' - \hat{\mathbf{ls}}_i|, \tag{7.5}$$

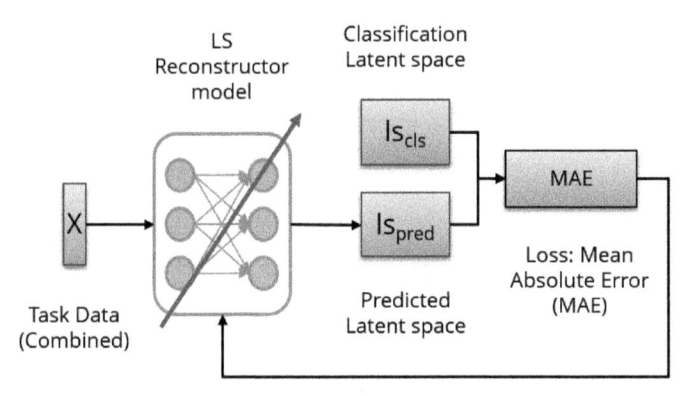

Task-Agnostic Inference Using Base–Child Classifiers   141

Fig. 7.9: Block diagram for the LS reconstructor. The LS reconstructor model is optimized using MAE loss, and the training set is formed by combining boundary points sample from all the tasks observed until now.

where $n(S_{\text{all}})$ is the cardinality of set S_{all}. Figure 7.9 describes the block diagram for the LS reconstructor model. The training completion for the LS reconstructor model also marks the end of the training phase for the BCC.

7.2.8 Child classifier

The child classifier model is a virtual extension to the LS reconstructor model and maps the output of the LS reconstructor (LS embedding/vector $\mathbf{ls}_i'$) model trained until task T_k to the classification head of the specified base classifier in order to obtain the prediction probabilities. No training is required for the child classifier, as it is a combination of two sets of pretrained models: the LS reconstructor and the base classifier's classification head. Figure 7.10 depicts the block diagram for the child classifier and describes how it is used to obtain predictions for test samples. Formally, to perform classification for any test sample of task $T_i \in \{T_1, T_2, \ldots, T_k\}$, the task T_i's classification head from its base classifier is attached on top of the LS reconstructor frozen model.

7.2.9 Automated task inference

The child classifier requires knowledge of task ID to select the respective classification head (from individual base classifiers).

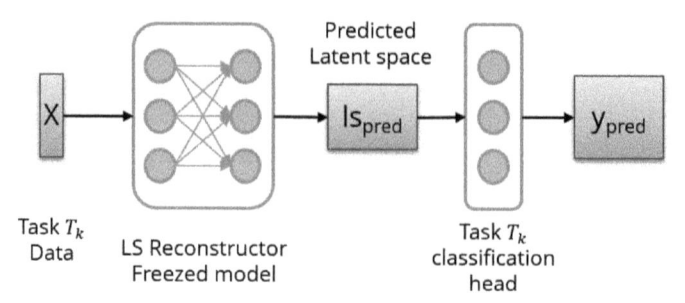

Fig. 7.10: Block diagram for the child classifier. This model appends the output layer of a specific base classifier (classification head) on top of the frozen LS reconstructor model to perform inference.

BCCs achieve automatic *task inference* (TI) using the reference points that were used for training the base classifiers. The distance between the current test sample's LS vector $\mathbf{ls}_i'$ and all the reference points in $R_{T_1}, R_{T_2}, \ldots, R_{T_k}$ is computed, and the task corresponding to the least distance is assigned as the inferred task ID. Let d_{T_k} be the least distance from $\mathbf{ls}_i'$ for task T_k:

$$d_{T_k} = \min\left\{\left\|R_{T_k}[j] - \mathbf{ls}_i'\right\|_2 : \forall j \in \{0, 1, \ldots, N_{T_k} - 1\}\right\} \quad (7.6)$$

$$id = \arg\min_r \{d_{T_r}\}_{r=1}^{k}, \quad (7.7)$$

where id is the inferred task ID. Since the task ID can be inferred at test time in BCCs, the method enables *task-agnostic* inference.

7.2.10 CL using BCCs

The key components of BCC and their descriptions have been covered in the above sections. It is now time to merge all the discussed units/modules to describe the working of BCCs under a CL scenario. The overall method was briefly reported in Section 7.2.2, and the entire procedure (training and testing phases) was illustrated in Fig. 7.1.

For a given task T_k (from a sequence of tasks), first train a base classifier using the current tasks' training data and store the boundary points S_{T_k} and classification head in memory. Update S_{all} by combining the boundary samples S_{T_k} for task T_k with previous tasks' boundary samples. Finally, train the LS reconstructor model using S_{all}, which concludes the training phase for BCC. The overall training algorithm for BCC is described in Algorithm 7.2. To perform

Algorithm 7.2 CL using BCCs: Training phase.

Require: $\{T_1, T_2, \ldots, T_k, \ldots, T_N\}$: A sequence of N classification tasks

Require: $D_{T_k} = \{(\mathbf{X}_1, y_1), (\mathbf{X}_2, y_2), \ldots, (\mathbf{X}_{N_k}, y_{N_k})\}$: For each task T_k, set of N_k training samples, with N_{T_k} class labels, where $\forall y_i \in \{0, 1, \ldots, N_{T_k} - 1\}$ and $\forall k \in \{0, 1, \ldots, N\}$

Require: Hyperparameters for each task T_k, such as (w_1, w_2): weights for the base classifier training loss, p: the fraction of boundary point samples selected per class

1: Let $f_{T_1}(\mathbf{x}; \boldsymbol{\theta})$ be the continual LS Reconstructor model, initialized with random weight $\boldsymbol{\theta}$
2: Obtain reference points $(R_{T_1}, R_{T_2}, \ldots, R_{T_N})$ corresponding to each task in the sequence
3: **for** $k \in \{0, 1, \ldots, N\}$ **do**
4:     Let $g_{T_k}(\mathbf{x}; \boldsymbol{\phi}_{T_k}) = h_{T_k}(e_{T_k}(\mathbf{x}; \boldsymbol{\alpha}_{T_k}); \boldsymbol{\omega}_{T_k})$ be Base classifier for task T_k, where $e_{T_k}(\mathbf{x}; \boldsymbol{\alpha}_{T_k})$ is the encoder network, $h_{T_k}(\mathbf{z}; \boldsymbol{\omega}_{T_k})$ is the classification head and $\boldsymbol{\phi}_{T_k} = (\boldsymbol{\alpha}_{T_k}, \boldsymbol{\omega}_{T_k})$ are initialized with random weights
5:     $L_{\text{weighted}} = w_1.L_{\text{CE}} + w_2.L_{\text{MAE}}$ (Refer Eq. (7.1), (7.2), and (7.3))
6:     Minimize L_{weighted} loss to train base classifier on training set of task T_k (D_{T_k}) to obtain learned parameters $\boldsymbol{\phi}_{T_k}^* = (\boldsymbol{\alpha}_{T_k}^*, \boldsymbol{\omega}_{T_k}^*)$
7:     Construct $Q_{T_k} = \left\{(\mathbf{X}_i, \hat{\mathbf{ls}}_i, y_i)\right\}_{i=1}^{N_{T_k}}$ for task T_k samples, where $\hat{\mathbf{ls}}_i = e_{T_k}(\mathbf{X}_i; \boldsymbol{\alpha}_{T_k}^*)$
8:     $S_{T_k} \leftarrow$ Algorithm 7.1 (Q_{T_k}, p)
9:     Store S_{T_k} and $h_{T_k}(.; \boldsymbol{\omega}_{T_k}^*)$ in memory
10:     **if** $k > 1$ **then**
11:         $S_{\text{all}} = \bigcup_{i=1}^{k} S_{T_i}$
12:         $L_{\text{rec}} = \frac{1}{n(S_{\text{all}})} \sum_{i=1}^{n(S_{\text{all}})} |\mathbf{ls}_i' - \hat{\mathbf{ls}}_i|$ (Refer Eq. (7.5))
13:         Let $f_{T_{k-1}}(.; \boldsymbol{\theta})$ be the trained LS reconstructor model after observing task T_{k-1} and $\boldsymbol{\theta}$ be the trained (or initialized) weights
14:         Minimize L_{rec} loss on training set S_{all} to obtain updated LS reconstructor model $f_{T_k}(.; \boldsymbol{\theta})$
15:     **end if**
16: **end for**
17: **return** $f_{T_N}(., \boldsymbol{\theta})$, $\{S_{T_k}\}_{k=1}^{N}$, $\{h_{T_k}(.; \boldsymbol{\omega}_{T_k}^*)\}_{k=1}^{N}$, $\{R_{T_k}\}_{k=1}^{N}$

Algorithm 7.3 CL using BCCs: Testing phase

Require: $\mathbf{X}_{\text{test}}$: Test sample
Require: $f_{T_N}(., \boldsymbol{\theta})$: LS reconstructor model trained up to task N
Require: $\{h_{T_k}(.; \boldsymbol{\omega}^*_{T_k})\}_{k=1}^{N}$: Classification heads for tasks $\{T_k\}_{k=1}^{N}$
Require: $\{R_{T_k}\}_{k=1}^{N}$: Reference points for tasks $\{T_k\}_{k=1}^{N}$
1: $\text{ls}'_{\text{test}} \leftarrow f_{T_N}(\mathbf{X}_{\text{test}}, \boldsymbol{\theta})$
2: **for** $k \in \{1, 2, \ldots, N\}$ **do**
3: $\quad d_{T_k} \leftarrow \min \{\|R_{T_k}[j] - \text{ls}'_{\text{test}}\|_2 :$
$\quad \forall j \in \{0, 1, \ldots, N_{T_k} - 1\}\}$
4: **end for**
5: $id \leftarrow \arg\min_k \{d_{T_k}\}_{k=1}^{N}$ (Refer Eq. (7.6) and (7.7))
6: $y'_{\text{test}} \leftarrow h_{T_{\text{id}}}(\text{ls}'_{\text{test}}; \boldsymbol{\omega}^*_{\text{id}})$
7: **return** y'_{test}
----

evaluation on test data for different tasks, first infer the task ID and then combine the LS reconstructor model with the inferred task ID's classification head to obtain the child classifier and the prediction for the test sample. The overall testing phase for BCC is provided in Algorithm 7.3.

7.3 Experimental Results and Discussion

This section covers the experiments performed on BCC to show the effectiveness of the method and provides detailed results and an in-depth discussion. Section 7.3.1 presents the datasets used for various experiments. Section 7.3.2 describes the evaluation metrics to benchmark the performance of BCC and a comparison of the baseline methods. Section 7.3.3 provides network architecture and training details for base classifiers and the LS reconstructor. The main results for BCC and its comparison with other baseline methods for multiple experiments are covered in Section 7.3.4. Sections 7.3.5 and 7.3.6 provide comparisons for different sampling methods (described in Section 7.2.6) and different sample sizes used for BCC, respectively.

7.3.1 Datasets and experimental setup

The experiments are performed on two standard datasets: MINST and Cifar10. Since the method needs to be evaluated for task-incremental continual learning of classification tasks, the following two experiments are designed:

- Split-Cifar10 (a homogeneous dataset): This experiment splits the Cifar10 dataset (consisting of 10 classes) into five classification tasks (T_1, T_2, ..., T_5), each comprising two classes (binary classification for each task). It is called a homogeneous dataset (experiment) as all the classification tasks belong to the same dataset (the source domain).
- Cifar10–MNIST (a heterogeneous dataset): This experiment comprises two tasks (T_1, T_2) over different domains: Cifar10 (10 class classification), followed by MNIST (10 class classification) classification tasks. Since the data for the two tasks belong to different data domains, it is a heterogeneous dataset (experiment).

7.3.2 Evaluation metrics and baseline methods

The experiments, Split-Cifar10 and Cifar10–MNIST, are evaluated using standard CL metrics, namely average accuracy (ACC) and backward transfer (BWT) (Lopez-Paz and Ranzato, 2017). ACC measures the average performance of the CL model on all the observed tasks. BWT measures the amount of forgetting incurred by the CL model when learning newer tasks. A large negative value for BWT indicates extreme forgetting and signifies *catastrophic forgetting*. Higher values for ACC and BWT indicate better performance by the CL model. Consider the test sets for N classification tasks, and after observing each task T_k, the CL model is evaluated on all the observed tasks T_1, T_2, ..., T_k. This gives rise to a matrix (table) $U \in \Re^{N \times N}$, where $U[i, j]$ indicates the test classification accuracy of the CL model on task T_j after observing the data from task T_i. For this study, the CL model is not evaluated on unseen tasks; hence, the entries in the upper triangle of U are not defined (indicated with "−" in the experimental results). Then, ACC and BWT are defined as follows:

$$\text{ACC} = \frac{1}{N} \sum_{i=1}^{N} U[N, i], \tag{7.8}$$

$$\text{BWT} = \frac{1}{N-1} \sum_{i=1}^{N} \left(U[N, i] - U[i, i] \right). \tag{7.9}$$

The experiments compared the performance of BCC on the above-described metrics on a variety of CL baseline methods, covering both regularization-based and replay-based approaches. The regularization-based methods compared were elastic weight consolidation (EWC) (Kirkpatrick et al., 2017), synaptic intelligence (SI) (Zenke et al., 2017), and learning without forgetting (LwF) (Li and Hoiem, 2018). In EWC, the goal is to remember old tasks while learning current task weights by imposing a penalty on changing old tasks' important weights. SI motivates the idea of intelligent synapses (from biological neurons), which accumulate task-relevant information over time and rapidly restore new knowledge without forgetting the old tasks. LwF uses the current task data to preserve the outputs of the CL model (trained until the previous tasks) for past tasks, acting as a regularizer when learning the current task. The compared replay-based methods were generative replay (GR) with variational autoencoders (GR with VAE) and averaged gradient episodic memory (A-GEM) (Chaudhry et al., 2018b). Deep generative replay (DGR) (Shin et al., 2017) uses a deep generative model, a generative adversarial network (GAN), learned over past task data, and replays samples when learning the current task. For this study, a variant of this method, in which GR is performed via a VAE model is considered. The other functions remain the same except the change in generative model from GAN to VAE. GEM (Lopez-Paz and Ranzato, 2017) uses episodic memory to avoid forgetting previous tasks and minimize negative backward transfer. A-GEM is an improved version of GEM and efficiently ensures that the average episodic memory loss over past tasks does not increase. In addition, BCC is also compared with traditional baselines, sequential fine-tuning (SFT) and joint training (JT). SFT trains incoming tasks in a transfer learning fashion, where no continual learning strategy is adopted and hence acts as *lower bound* to CL methods. JT trains all the seen tasks together and has access to past task data at each step, which is not the case for CL and hence acts as a potential *upper bound* for CL approaches.

7.3.3 Network architecture and training details

This section describes the network architectures for various modules of BCC, such as base classifiers and the LS reconstructor, and other hyperparameters, such as reference points, loss weights, and optimizers, used during the training phase.

Reference points for experiments were chosen using the strategy described in Section 7.2.3. For the Cifar10–MNIST experiment, the two-task scenario method was used by partitioning the LS into all positive and negative regions and randomly assigning +1 (for MNIST) and −1 (for Cifar10) to one of the LS dimensions (and the remaining LS dimensions to 0) for respective tasks. For the Split-Cifar10 experiment, the reference points chosen followed a variation of the generic method described in Section 7.2.3. Since each task is a two-class classification, reference points for respective classes are assumed to lie on hyperspheres of radii 1 and 2 units, ensuring inter-class separation. Inter-task separation in enforced by choosing five random LS dimensions (corresponding to five tasks) and assigning 1 or 2 for those dimensions and 0 for the remaining dimensions.

For both experiments, Split-Cifar10 and Cifar10–MNIST, the base classifiers and LS reconstructor have the same convolution neural network (CNN) architecture, except for the classification head layer in the base classifiers. It consisted of five convolution layers (all layers except the first have a stride of 2), batch-norm and ReLU activation layers, followed by two dense (linear) layers and a softmax layer for output (with 2 and 10 nodes per task for Split-Cifar10 and Cifar10–MNIST, respectively). In this study, the network capacity for both the base classifier and the LS reconstructor was the same; however, this is not a strict requirement. The LS reconstructor model can have lower or higher parameters depending on the complexity of the CL problem at hand. The LS dimension s and sampling fraction p for both experiments are 128 and 0.1, respectively. Loss weights (w_1, w_2) between cross-entropy and clustering losses for the base classifier training of Split-Cifar10 and Cifar10–MNIST were empirically set as (0.1, 1) and (1, 1), respectively. The reported values were obtained after performing a hyperparameter search and observing test performance. The base classifiers for individual tasks of Split-Cifar10 and Cifar10–MNIST used batch sizes of 16 and 256, respectively, with 100 epochs and 10% training samples used for validation. The LS reconstructor for Split-Cifar10 was trained for 150 epochs with a batch size of 16, whereas the Cifar10–MNIST LS reconstructor was trained using a batch size of 256 for 250 epochs. The Adam optimizer (with a learning rate of 0.001) was used for training both base classifiers and LS reconstructor models for both experiments. As mentioned earlier, the sampling fraction for storing boundary point samples was 0.1; however, in order to boost the training performance of the LS reconstructor model, additional augmentations (9× of sample size to obtain the same training set size as that of the original datasets) based on geometric operations, such as rotation (up to 10 deg), width shift (up to 0.1 fraction of image width) and height shift (up to 0.1 fraction of image height) were used. For a fair comparison, the same network architectures (as used for base classifiers) were used when training the other discussed baseline methods. The above network architectures, hyperparameters and training details for BCC were used to perform the experiments, and their results (averaged over three runs) are reported in Section 7.3.4.

Sections 7.3.5 and 7.3.6 provide additional results on various scenarios performed only on BCC to understand the effect of sampling techniques and sampling sizes. For this portion of the study, a slight variation on the above-reported network architecture is used. The CNN architecture for both the base classifier and the LS reconstructor consists of six convolution layers, three pooling layers (after every couple of convolution layers) and leaky ReLU activation layers. The convolution block is followed by a 128 node linear layer (with leaky ReLU activation) and a softmax layer as output. The remaining training details, such as batch size, optimizers, training epochs, loss weights and reference points, are the same. The results for varying sampling fraction p across different sampling methods are reported. All experiments reported in this chapter were executed on a workstation with 64 GB of RAM and a GTX 2080Ti GPU.

7.3.4 Main results and discussion

The experiments are performed for two scenarios: "with task ID," where the task ID is provided during testing, and "without task ID," where task ID is required to be inferred to perform inference. BCC is designed to perform automatic task inference; however, other baseline methods do not possess a direct way to perform task inference. A well-accepted strategy is to perform the classification across all the heads and then choose the predicted class corresponding to the highest predicted probability (softmax score). Hence, for the "without task ID" scenario, this approach is used to perform inference for other baseline methods. This method of inference is also performed for BCC, and the method is dubbed "BCC - argmax" to differentiate it from "BCC - TI," where automatic task inference is used.

Table 7.1 provides the results for the Split-Cifar10 experiment for both scenarios of with and without task ID. When comparing ACC (Eq. (7.8)), BCC - TI outperforms all compared methods, achieving a score of 73.28%. Other CL methods report ACC scores of less than 25%, including BCC - argmax, showing that the task inference using argmax is quite poor as compared to BCC's task inference method using reference points. Even JT shows a lower score than BCC - TI, which could be attributed to incorrect task inference, as, when

Table 7.1: Results for the Split-Cifar10 experiment. BCC outperforms other methods for the "without task ID" scenario.

	Without task ID		With task ID	
Method	ACC	BWT	ACC	BWT
SFT	0.1818	−0.8669	0.6258	−0.3130
JT	0.5984	−0.1274	0.8697	0.0003
EWC (Kirkpatrick et al., 2017)	0.1799	−0.8567	0.8004	−0.0615
SI (Zenke et al., 2017)	0.1795	−0.8552	0.8217	−0.0419
LwF (Li and Hoiem, 2018)	0.1842	**−0.0003**	**0.8549**	**−0.0132**
GR with VAE	0.1797	−0.7769	0.7521	−0.1211
A-GEM (Chaudhry et al., 2018b)	0.2454	−0.2834	0.8170	−0.0901
BCC - argmax	0.2196	−0.3122	0.7416	−0.1215
BCC - TI	**0.7328**	−0.1071	0.7416	−0.1215

Note: The bold terms represent the best scores. Higher value for ACC and lower value for BWT are better.

compared to the "with task ID" case, it improves to 86.97% (beating all CL methods, as expected). When comparing the BWT score, LwF achieves the best performance of −0.0003. However, LwF is not the best CL model, as its corresponding ACC indicates that the LwF model focused too much on remembering past tasks, which hindered its learning of future tasks. Here again, the BWT score of JT is poorer than that of BCC - TI, which is attributed to poor task inference, as, for the "with task ID" scenario, it improves to a positive backward transfer of 0.0003. This clearly demonstrates the efficacy of BCC, as it beats all compared methods in terms of ACC and BWT. When comparing results in the "with task ID" scenario, it can be observed that all the methods boost their respective scores. However, the improvement for BCC - TI is marginal, from 73.28% to 74.16%, as compared to other methods. All other methods were designed to perform for the "with task ID" scenario, and hence, it is expected from them to perform better, whereas BCC is designed with in-built task inference and is more suited for the "without task ID" scenarios. LwF achieves the best results for both ACC and BWT in the "with task ID" case. Despite that, the performance of BCC is not as poor as that of SFT and is comparable to GR with VAE.

Figures 7.11 and 7.12 depict the ACC and BWT scores computed after observing each task of the Split-Cifar10 experiment for the "without task ID" scenario, respectively, across all discussed methods. For BWT, the axis starts from task 2, as there is no forgetting measure for a single task. When comparing ACC scores, BCC - TI

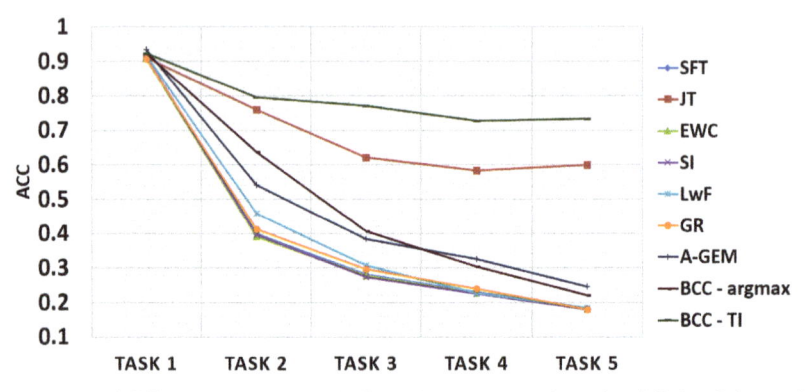

Fig. 7.11: ACC scores computed after observing each task of Split-Cifar10 for the "without Task ID" scenario, across all discussed methods.

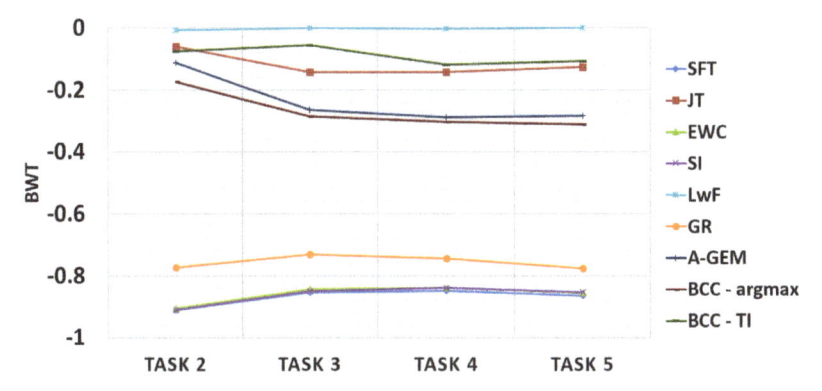

Fig. 7.12: BWT scores computed after observing each task of Split-Cifar10 for the "without Task ID" scenario, across all discussed methods.

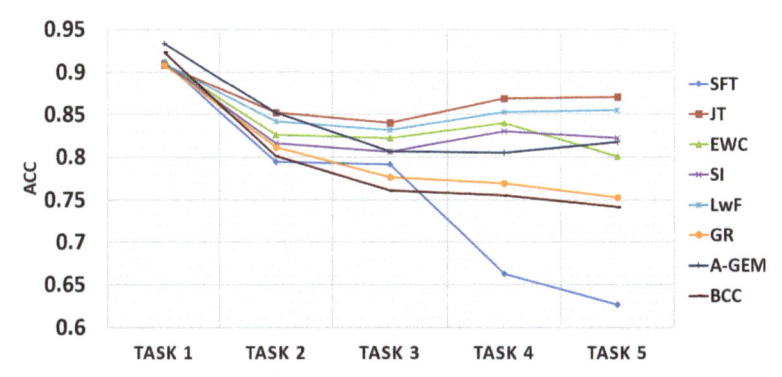

Fig. 7.13: ACC scores computed after observing each task of Split-Cifar10 for the "with task ID" scenario, across all discussed methods.

is consistently better than other methods at each task step, whereas SFT, EWC and SI are among the worst performers. When comparing BWT scores across tasks, LwF follows an almost horizontal line with a score of 0. BCC - TI also shows a consistent BWT score across tasks, indicating that the method is resilient to forgetting. Similarly, Figs. 7.13 and 7.14 demonstrate the ACC and BWT scores computed after observing each task of Split-Cifar10 for the "with task ID" scenario, respectively, across all discussed methods. Observing ACC scores, JT clearly outperforms other CL methods, followed by LwF, and SFT performs the worst. BCC follows a similar path to that of GR with VAE. From the BWT plot, JT consistently shows positive backward transfer, followed by LwF, which also maintains a small

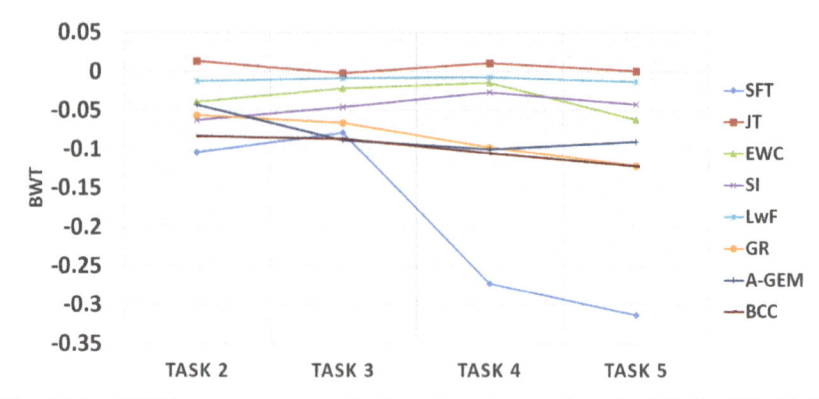

Fig. 7.14: BWT scores computed after observing each task of Split-Cifar10 for the "with task ID" scenario, across all discussed methods.

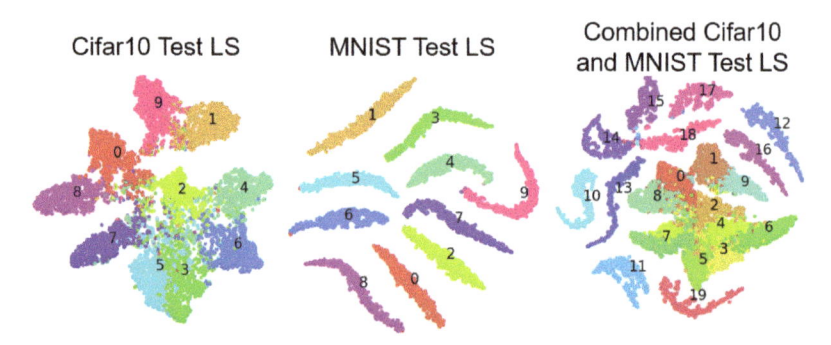

Fig. 7.15: t-SNE visualization for Cifar10 (left) and MNIST (center) classification obtained using respective base classifiers on corresponding test sets. t-SNE visualization for combined Cifar10 (indices 0–9) and MNIST (indices 10–19) classification obtained from the child classifier (right).

negative BWT score across tasks. There is a small drop observed for BCC, as the BWT score goes from -0.0830 to -0.1215.

For the Cifar10–MNIST experiment, individual base classifiers trained on Cifar10 and MNIST achieved an accuracy of 66.69% and 99.23% on respective test sets. The performance of the child classifier after observing both tasks is 49.79% for Cifar10 and 96.90% for MNIST, leading to an ACC score of 73.35 for the "with task ID" scenario. Figure 7.15 depicts the t-SNE visualizations of latent space for the Cifar10 base classification, the MNIST base classification and the joint Cifar10–MNIST child classification. It can be observed that the MNIST classes are able to coexist within the same latent space as that of Cifar10 and also preserve inter-class separation (separation within MNIST classes) as well as inter-task separation (separation with all of Cifar10 classes). This shows that the child classifier obtained using the BCC approach is able to retain the knowledge from individual base classifiers and inhibits forgetting of previous tasks.

Table 7.2 provides the results for the Cifar10–MNIST experiment for both scenarios, with and without task ID. When comparing ACC scores for the "without task ID" scenario, both the BCC versions (BCC - argmax and BCC - TI) achieve state-of-the-art performance, beating all other CL baselines. BCC - TI and BCC - argmax obtain ACC scores of 76.88% and 73.35%, respectively. BCC - argmax attains a slightly better score than BCC - TI, this could be attributed to the presence of contrasting output probabilities for Cifar10 and MNIST, leading to better task inference. When comparing BWT scores, BCC - argmax again outperforms other CL baseline methods, achieving a score of −0.0978. Upon comparing the ACC scores for the "with task ID" scenario, all the methods observe an improvement as compared to their "without

Table 7.2: Results for the Cifar10–MNIST experiment. BCC outperforms other methods for the "without task ID" scenario.

Method	Without task ID ACC	Without task ID BWT	With task ID ACC	With task ID BWT
SFT	0.4941	−0.6165	0.6020	−0.4007
JT	0.8115	0.0306	0.8122	0.0046
EWC (Kirkpatrick et al., 2017)	0.5027	−0.5553	0.6650	−0.2324
SI (Zenke et al., 2017)	0.5060	−0.5840	0.6342	−0.2957
LwF (Li and Hoiem, 2018)	0.3177	−0.2121	**0.7726**	−0.1487
GR with VAE	0.6997	−0.1582	0.7412	−0.0998
A-GEM (Chaudhry et al., 2018b)	0.6401	−0.3371	0.7527	−0.1336
BCC - argmax	**0.7688**	−0.0978	0.7697	**−0.0965**
BCC - TI	**0.7335**	−0.1690	0.7697	**−0.0965**

Note: The bold terms represent the best scores. Higher value for ACC and lower value for BWT are better.

task ID" counterparts. LwF attains the best ACC score of 77.26%, followed by BCC, which achieves 76.97%. The gap between the two ACC scores is marginal, and given that BCC obtains the best BWT score of -0.0965, it is safe to say that BCC outperforms other CL baselines for the "with task ID" scenario as well. The observations for the Cifar10–MNIST experiment are consistent with those of Split-Cifar10 for the "without task ID" scenario but better when compared to the "with task ID" case.

In general, the baseline methods perform well in the presence of a task identifier. However, in its absence, their ACC drops significantly from 57.16% (= 81.70 − 24.54) for A-GEM to 67.07% (= 85.49 − 18.42) for LwF, for the Split-Cifar10 experiment and from 4.15% (= 74.12 − 69.97) for GR with VAE to 45.49 (= 77.26 − 31.77)% for LwF, for the Cifar10–MNIST experiment (refer Tables 7.1 and 7.2). In contrast, BCC showcases comparable performance in the presence of task ID, whereas without task ID, it outperforms all other baselines, by 48.74% (= 73.28 − 24.54) and 6.91% (= 76.88 − 69.97) with homogeneous (Split-Cifar10) and heterogeneous (Cifar10–MNIST) tasks, respectively. BCC - TI and BCC - argmax outperform all CL baseline methods for the "without task ID" scenario on Split-Cifar10 and Cifar10–MNIST, respectively.

7.3.5 Comparison for different sampling methods

Sampling boundary points is a key component in BCC, and for the results discussed in Section 7.3.4, the farthest from each point sampling strategy has been used. In this section, this technique is compared with other sampling methods described in Section 7.2.6, and its effectiveness is demonstrated.

For this study, all the experiments performed used task ID during test time since, as previously observed from Section 7.3.4, there was not much difference between the two scenarios (with and without task ID) for BCC. Table 7.3 provides the results for various sampling methods used with BCC for a sampling fraction of $p = 0.1$ (or a sampling percentage of 10%) on the Cifar10–MNIST experiment. The columns "Cifar10" and "MNIST" indicate the test accuracy scores for Cifar10 and MNIST, respectively, after observing both tasks. Some methods, such as disjoint univariate combinations sampling, outermost cluster sampling, and t-SNE

Table 7.3: Results for various sampling methods used with BCC for the Cifar10–MNIST experiment. The sampling fraction p is set at 0.1, or 10%, for all methods.

Sampling method	Cifar10	MNIST	ACC	BWT
Base classifier	0.8184	0.9964	—	—
Random	0.7243	0.9964	0.8520	−0.0941
Farthest distance	0.7040	0.8993	0.8017	−0.1144
Disjoint univariate combinations	0.7166	0.9633	0.8400	−0.1018
Outermost cluster	**0.7720**	0.3449	0.5585	**−0.0464**
Low-density Gaussian distribution	0.6975	0.9823	0.8399	−0.1209
t-SNE proj.-based concave hull	0.7028	**0.9841**	0.8435	−0.1156
Farthest from each point	0.7323	0.9802	**0.8563**	−0.0861

Note: The bold terms represent the best scores. Higher value for ACC and lower value for BWT are better.

projections-based concave hull sampling, do not exactly meet the 10% sampling percentage mark but exhibit close-by numbers. This is because the number of samples selected for these methods is controlled by other specific hyperparameters and not governed directly by the sampling percentage parameter. Also, two methods, univariate combinations sampling and t-SNE projections-based convex hull sampling, are not present in the above table, as samples generated using these techniques were significantly less than the 10% quota. More details about such methods are provided in Section 7.3.6. The first row of the table is special and does not represent any sampling method; rather, it specifies the scores corresponding to the respective base classifiers of Cifar10 and MNIST. The middle block depicts the scores for various sampling techniques, and the final row indicates the score for the farthest from each point sampling method. When comparing individual test scores for Cifar10 and MNIST, the outermost cluster sampling and t-SNE projections-based concave hull sampling outperform other methods, respectively. However, outermost cluster sampling does perform poorly for MNIST, leading to a poor ACC score as well. Similarly, the high score of BWT can be justified, as the method completely focused on remembering Cifar10 and did not learn MNIST and, in doing so, achieved a good BWT score. By ignoring outermost clustering, the method that excels is farthest from each point sampling, achieving the ACC and BWT scores of 85.63% and −0.0861, respectively. Comparing ACC across

Table 7.4: Results for various sampling methods used with BCC for the Cifar10–MNIST experiment. The sampling fraction p is set to 0.05 or 5% for all methods.

Sampling method	Cifar10	MNIST	ACC	BWT
Base classifier	0.8184	0.9964	—	—
Random	0.6434	**0.9755**	0.8095	−0.1750
Farthest distance	0.5962	0.7841	0.6902	−0.2222
Univariate combinations	0.6627	0.9693	0.8160	−0.1557
Low-density Gaussian distribution	0.6265	0.9709	0.7987	−0.1919
t-SNE proj.-based concave hull	0.6422	0.9696	0.8059	−0.1762
Farthest from each point	**0.6681**	0.9681	**0.8181**	**−0.1503**

Note: The bold terms represent the best scores. Higher value for ACC and lower value for BWT are better.

different methods, the gap between scores is not high (0.46%), ranging from 80.17% for farthest distance sampling to 85.63% for farthest from each point sampling (ignoring outermost cluster sampling with a 55.85% ACC score). It indicates that the variation between different sampling methods is not drastic; nevertheless, the farthest from each point sampling method still outperforms the other discussed approaches.

Similarly, Table 7.4 shows the results for the same experiment using 0.05 (or 5%) as the sampling fraction (or percentage). There are some omissions, such as outermost cluster sampling and disjoint univariate combinations sampling, as the exact number of samples (5%) required for a fair comparison was not available using these methods. Univariate combination sampling is able to participate in this comparison study. More details related to them are provided in Section 7.3.6. The farthest from each point and random sampling methods outperform for Cifar10 and MNIST test classifications, respectively. Considering the ACC and BWT metrics, farthest from each point sampling achieves the best score compared to other methods. The overall performance comparison of the farthest from each point sampling method for both 5% and 10% justifies its use as a boundary points sampling method with BCC.

7.3.6 *Comparison for different sample sizes*

This section presents a discussion on using different sampling sizes for sampling methods and their impact on BCC's performance.

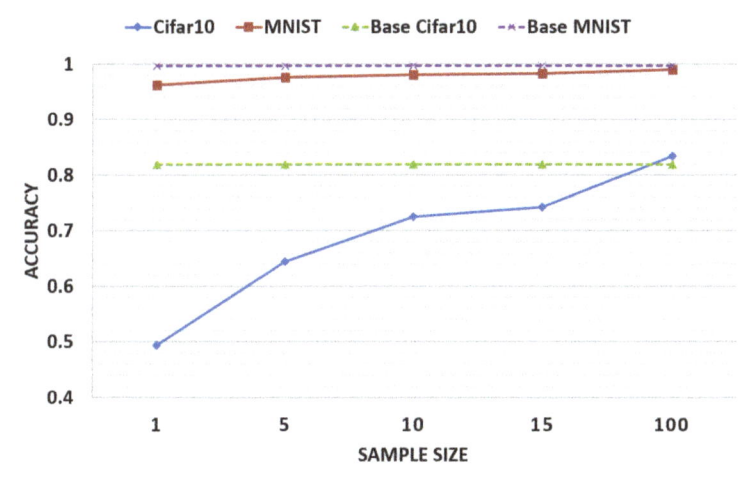

Fig. 7.16: Test accuracy scores for Cifar10 and MNIST after observing both tasks, across different sampling sizes using random sampling.

Figure 7.16 depicts the plot for test accuracy scores for Cifar10 and MNIST after observing both tasks for different sample sizes, ranging from 1% to 100%, using random sampling, along with the base classifiers' performance of Cifar10 (light green curve) and MNIST (purple curve). It can be observed that even using just 1% samples, the Cifar10 and MNIST accuracy scores are 49.30% and 96.22%, respectively, which are quite impressive. With an increase in sample size, the margin of improvement in MNIST is small, ranging from 96.22% for 1% samples to 98.89% for 100% samples (all samples selected), whereas Cifar10 makes a huge leap from 49.30% to 83.30%. This suggests that even a small fraction of MNIST samples randomly selected works achieve good performance, whereas for Cifar10, a relatively larger fraction (e.g., 10% or 15%) is required. This could be due to different image and classification complexities for both datasets. A similar comparison is provided using farthest distance sampling in Fig. 7.17. One contrasting difference in the performance of small sample sizes (especially for MNIST) is observed when compared to the random sampling plot. The corresponding accuracy scores for the same sample sizes are quite lower for the farthest distance method. This could be due to samples being selected from the same/similar region(s) of the LS, leading to mode collapse (a lack of diversity in selected samples) and poor performance on the test set.

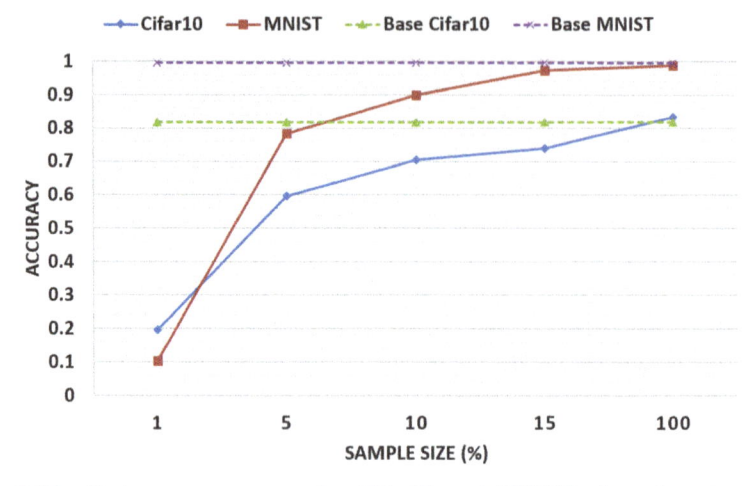

Fig. 7.17: Test accuracy scores for Cifar10 and MNIST after observing both tasks, across different sampling sizes using farthest distance sampling.

Table 7.5: Results for different sample sizes when using the disjoint univariate combinations sampling method with BCC for the Cifar10–MNIST experiment.

Selection type	Clusters	Size (%)	Cifar10	MNIST	ACC	BWT
min, max	1	2.65	0.5772	0.9162	0.7467	−0.2412
min, max, mean	1	4.92	0.6627	0.9693	0.8160	−0.1557
min, max, mean	2	9.10	0.7166	0.9633	0.8400	−0.1018
min, max, mean	3	12.73	0.7366	0.9722	0.8544	−0.0818
min, max, mean	4	15.82	0.7571	0.9708	0.8640	−0.0613
min, max, mean	5	18.44	0.7613	0.9690	0.8652	−0.0571
min, max, mean	6	20.84	0.7677	0.9808	0.8743	−0.0507

However, once enough samples are present, the performance improves for Cifar10 and MNIST and attains similar (still inferior) scores as that random sampling.

Table 7.5 displays the results for different parameters used with disjoint univariate combinations sampling, leading to different sample sizes. As discussed in Section 7.2.6, disjoint univariate combinations sampling is influenced by selection type, which includes minimum, maximum and mean selection, and the number of disjoint clusters.

When the number of clusters is 1, it becomes univariate combinations sampling. The column "size (%)" denotes the number of samples selected based on values for these two parameters. Columns "Cifar10" and "MNIST" denote the test accuracy scores obtained using the BCC model after observing both tasks, and "ACC" and "BWT" represent the respective CL metrics. The first row block shows the results for univariate combinations sampling (as the number of clusters = 1), and the second block shows the results for disjoint univariate combinations sampling by varying the number of clusters from 2 to 6. As the selection type or number of clusters are increased, the sampling percentage increases, leading to consistent improved performance of both Cifar10 and MNIST, thereby resulting in boosted ACC and BWT scores.

Similarly, Table 7.6 provides the results for various parameter settings used with t-SNE projections-based convex/concave sampling. Here, one of the parameters is whether to use a convex or concave hull on t-SNE projections to obtain selections. If the method is executed a single time, the selected boundary will be thin (less number of samples selected). The same approach can be repeated for multiple iterations (column "iter." in the table) in order to obtain a thick boundary, leading to an increase in selected samples. The first block (row) indicates the results for samples selected using the convex hull criterion. As reported, it corresponds to selecting a very low percentage (0.44%) of samples. The second block shows the results for the concave hull strategy across increasing iterations from 1 to 4. As evidenced by other sampling methods, with an increase in sampling size, the performance of BCC improves. The same is true for

Table 7.6: Results for different sample sizes when using the t-SNE projections-based convex/concave hull sampling method with BCC for the Cifar10–MNIST experiment.

Hull type	Iter.	Size (%)	Cifar10	MNIST	ACC	BWT
Convex	1	0.44	0.4414	0.9097	0.6756	−0.3770
Concave	1	2.53	0.5609	0.9742	0.7676	−0.2575
Concave	2	5.20	0.6422	0.9696	0.8059	−0.1762
Concave	3	7.92	0.6847	0.9819	0.8333	−0.1337
Concave	4	10.70	0.7028	0.9841	0.8435	−0.1156

t-SNE projections-based convex/concave sampling, and it achieves comparable performance to other methods for similar sampling percentages.

From the above results, it is evident that the performance of BCC improves with the increase in the number of selected samples. However, in a CL setting, only a small fraction of samples can be stored. One simple way to increase this number is by using augmentations for the stored samples. Hence, even though only a limited percentage of samples are stored, using geometric operations, such as image rotation, width shift and height shift on images, more images can be constructed on the fly, and only LS representations for these samples are needed to be stored in memory. Table 7.7 provides the results for the farthest from each point sampling method across different sample sizes. The columns "$p\%$," "aug." and "overall" represent the sample size selected using a method, additional augmentations constructed (as a multiple of selected samples) and overall samples (samples selected + augmentations) used to train the LS reconstructor model, respectively. The first block depicts the results for $p = 5\%$ and two cases: one with no augmentation and the other with 5× augmentations. When training BCC with additional samples obtained using augmentations, the test accuracy scores for both Cifar10 and MNIST improve, along with an improvement in ACC (from 81.81% to 84.65%) and BWT (from −0.1503 to −0.1116) scores. The middle block shows the results for $p = 10\%$ and augmentations ranging from 0× to 9×. For this case as well, there is a consistent increase

Table 7.7: Results for different sample sizes when using the farthest from each point sampling method with BCC for the Cifar10–MNIST experiment.

p(%)	Aug.	Overall (%)	Cifar10	MNIST	ACC	BWT
5	0×	5	0.6681	0.9681	0.8181	−0.1503
5	5×	30	0.7068	0.9862	0.8465	−0.1116
10	0×	10	0.7323	0.9802	0.8563	−0.0861
10	1×	20	0.7520	0.9886	0.8703	−0.0664
10	2×	30	0.7619	0.9953	0.8786	−0.0565
10	4×	50	0.7722	0.9955	0.8839	−0.0462
10	9×	100	0.7773	0.9954	0.8864	−0.0411
100	0×	100	0.8330	0.9889	0.9110	+0.0146

in all metrics. Specifically, there is a 3.01% and 0.0450 improvement in ACC and BWT scores, respectively. The last row denotes the performance of BCC when training using all the training set samples (for both tasks). Even though the number of samples for the $p = 100\%$ and $p = 10\% + 9\times$ augmentation cases is the same, the performance of the former is better than that of the latter, as the diversity present in the entire training set samples is very hard to capture using just the augmentations constructed using 10% samples. However, the gap in ACC scores for these two cases is only 2.46%, which is not that significant. Thus, this demonstrates that the farthest from each sampling method, along with augmentations, can achieve good performance on CL tasks.

7.4 Conclusion and Future Directions

In this chapter, a task-agnostic inference method for CL classification tasks is discussed, named BCCs. It comprises base–child hybrid networks for learning shared representations across tasks, allowing for effective coexistence and retention of learned knowledge and enabling intra- and inter-task separation using reference points. Along with the boundary point sampling method used in BCCs, an additional detailed study on sampling methods is also discussed. Experimental results for a task-agnostic setting showed that BCCs attain the best performance on both homogeneous (Split-Cifar10) and heterogeneous (Cifar10–MNIST) tasks with a minimal drop in accuracy when compared to task information-dependent methods. Extended studies performed on sampling approaches and the impact on sample size on BCC provided key insights on different sampling methods and established the effectiveness of the farthest from each point sampling method.

For future studies, one key area is to further expand the experimental study to more than five tasks. Since the selection of reference points is an essential step for BCCs, other methodologies can be explored. Another interesting direction is to reduce the number of boundary points sample selected (to be replayed later for the LS reconstructor model) from 5% or 10% to as low as 1% or 2% without compromising on performance.

© 2024 World Scientific Publishing Company
https://doi.org/10.1142/9789811286711_0008

Chapter 8

Flashcards for Knowledge Capture and Replay

Saisubramaniam Gopalakrishnan[*,‡], Pranshu Ranjan Singh[*,§],
Haytham M. Fayek[†,∥], Savitha Ramasamy[*,¶]
and ArulMurugan Ambikapathi[*,**]

[*]*Institute for Infocomm Research, Agency for Science,
Technology and Research (A*STAR), Singapore*
[†]*School of Computing Technologies, Royal Melbourne Institute of
Technology (RMIT) University, Melbourne, Australia*
[‡]*saisubramaniam147@gmail.com*
[§]*pranshu96ranjan@gmail.com*
[∥]*haytham.fayek@ieee.org*
[¶]*ramasamysa@i2r.a-star.edu.sg*
[**]*a.arulmurugan@gmail.com*

Abstract
Deep neural networks model data for a task or a sequence of tasks, where the knowledge extracted from the data is encoded in the parameters and representations of the network. Extraction and utilization of these representations are vital when data are no longer available in the future, especially in a continual learning scenario. Flashcards are visual representations that capture the encoded knowledge of a network as a recursive function of some predefined random image patterns. In a continual learning scenario, flashcards help prevent catastrophic forgetting by consolidating the knowledge of all the previous tasks. Flashcards are required to be constructed only before learning the subsequent task and

are hence independent of the number of tasks trained previously, making them task-agnostic. The efficacy of flashcards in capturing learned knowledge representation (as an alternative to the original data) is empirically validated on a variety of continual learning tasks: reconstruction, denoising and task-incremental classification, using several heterogeneous (varying background and complexity) benchmark datasets. Experimental evidence indicates that flashcards as a replay strategy are task-agnostic, perform better than generative replay and are on par with episodic replay without additional memory overhead.

8.1 Introduction

Deep neural networks (DNNs) are efficient function approximators for modeling data sampled from the underlying distribution of a single task (which may comprise one or more classes). The parameters (weights) and representations (activations/features) of DNNs are encoded with the knowledge extracted during the training of a particular task(s) using these data points. When data are no longer available in the future (e.g., in continual learning), it is imperative to extract and utilize these representations in order to retain a sense of the past. While DNNs have proven successful in multiple domains for single tasks (Krizhevsky *et al.*, 2012; Längkvist *et al.*, 2014; Devlin *et al.*, 2019), there exists a level of difficulty in extracting and reusing the knowledge of trained models for similar or different downstream scenarios (Bing, 2020). Standalone techniques, such as transfer learning (Zhuang *et al.*, 2021) and knowledge distillation (Gou *et al.*, 2020), facilitate representational or functional translation (Hinton *et al.*, 2015) from one model to another. However, they cannot construct a unified representation that evolves with changing data characteristics (due to class or concept drift). Gaining new knowledge through learning new tasks affects the retention of prior knowledge (Robins, 1995). The concept of continual learning (CL) (Thrun, 1996) was proposed as a strategy for such retention, and it is getting more attention (Parisi *et al.*, 2019). The core goal of CL is to make DNNs continuously learn from a sequence of tasks without catastrophically forgetting the previous sequences/tasks. The objective is to come up with a reliable method for capturing and repurposing knowledge from a learned model, which can then be applied in the context of CL. Knowledge captured from the past should be consolidated and

Flashcards for Knowledge Capture and Replay 165

preserved as some intelligent representation with low computational overhead and memory requirements.

This chapter introduces *flashcards* (Gopalakrishnan et al., 2020), a mechanism inspired by the brain (Xue et al., 2010; Danker and Anderson, 2010; Buch et al., 2021) to capture consolidated representations of learned knowledge from previous tasks. Specifically, flashcards are visual representations that *capture* the encoded knowledge of a network as a recursive function of predefined random image patterns. A schematic illustration is shown in Fig. 8.1. Recursively passing random images through a trained autoencoder (obtained at the end of step (a)) captures the representation with each pass and is then converted into flashcards (step (b)). To train

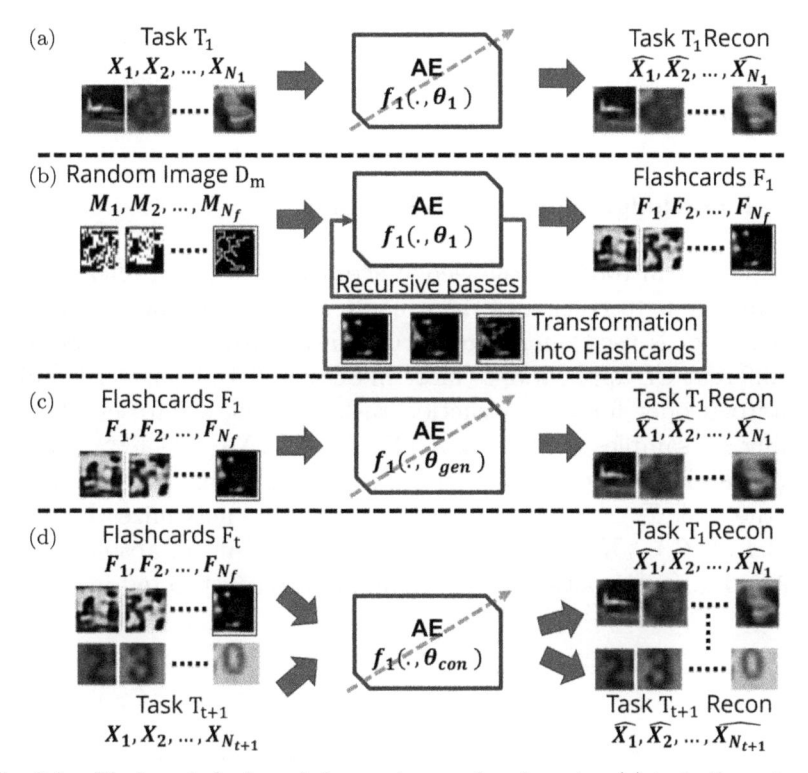

Fig. 8.1: Flashcards for knowledge capture and replay: step (a) train the autoencoder (AE) for Task T_1; step (b) construct flashcards from frozen AE using maze patterns via recursive passes; step (c) replay using flashcards on a new network to remember Task T_1; step (d) replay using flashcards F_t (containing consolidated knowledge from task T_1 to T_t) while training for Task T_{t+1}.

a new network from the obtained flashcards (single task), step (c) is used, and step (d) addresses the multi-task/continual setting.

Initially, flashcards' capabilities to capture representations of data (as seen by the network) are examined, and later, the functionality is extended for replay of several heterogeneous datasets while learning continuously. Flashcards are an effective way to prevent catastrophic forgetting in CL scenarios. They consolidate the knowledge of all the previous tasks and prevent catastrophic forgetting. The construction of flashcards is needed only before learning the next task, so they are agnostic to the number of tasks learned before. As an alternative to the original data, the efficacy of flashcards in tracking learned knowledge is demonstrated by empirical validation using several heterogeneous (varying background and complexity) benchmark datasets. According to experimental evidence, flashcards as a replay strategy perform better than generative replay, are on par with episodic replay and contain no additional memory overhead when compared to generative replay.

8.2 Knowledge Capture and Replay

This section introduces knowledge acquisition using flashcards and describes how to construct them using a trained autoencoder. The autoencoder is used *only* to construct flashcards, which can be used for different CL applications (more in Section 8.4). Consider training an autoencoder for reconstruction task T_t using a dataset D_t containing N_t samples, where $D_t = \{\mathbf{X}_1, \mathbf{X}_2, \ldots, \mathbf{X}_{N_t}\} \subset \Re^{k \times l \times c}$ is the set of training image samples with k rows, l columns, and c channels. For the task T_t, the autoencoder is trained to maximize the likelihood $P(\mathbf{X}|\boldsymbol{\theta}_t), \forall \mathbf{X} \in D_t$ using the conventional mean absolute error (MAE; Eq. (8.1)) between the original and reconstructed images:

$$\min_{\boldsymbol{\theta}_t} \frac{1}{N_t} \sum_{n=1}^{N_t} |\mathbf{X}_n - \widehat{\mathbf{X}}_n|, \qquad (8.1)$$

where $\boldsymbol{\theta}_t$ is the autoencoder network parameters (weights, biases, and batch norm parameters) for task T_t, $\mathbf{X}_n$ is a sample from the empirical data distribution of $P(X)$, and $\widehat{\mathbf{X}}_n = f_t(\mathbf{X}_n, \boldsymbol{\theta}_t)$ is the reconstructed sample. $|\mathbf{X}_n - \widehat{\mathbf{X}}_n|$ denotes the pixel-wise MAE between $\mathbf{X}_n$ and $\widehat{\mathbf{X}}_n$, and $f_t(\cdot)$ is the function approximated by the autoencoder to learn (reconstruct) task T_t. In the above conventional learning setup, the parameters θ_t of an autoencoder network aim to model the knowledge in the data D_t, such as the *shape, texture,* and *color* of the images.

Let the reconstruction error of the trained autoencoder be bounded by $[\epsilon_1, \epsilon_2]$, i.e.,

$$\epsilon_1 \leq |\mathbf{X}_n - \widehat{\mathbf{X}}_n| \leq \epsilon_2, \forall \mathbf{X}_n \in D_t, \mathbf{X}_n \sim P(\mathbf{X}). \quad (8.2)$$

Let $P(\mathbf{M})$ be a different but well-defined distribution in the same dimensional space as $P(\mathbf{X})$ (i.e., $\Re^{k \times l \times c}$), and $D_m = \{\mathbf{M}_1, \mathbf{M}_2, \ldots, \mathbf{M}_{N_f}\}$, where $\mathbf{M}_i \in \Re^{k \times l \times c}$, $\mathbf{M}_i \sim P(\mathbf{M}), i = 1, \ldots, N_f$. The output of the autoencoder for any $\mathbf{M}_i \in D_m$ is

$$\widehat{\mathbf{M}}_i = f_t(\mathbf{M}_i, \theta_t). \quad (8.3)$$

Since the autoencoder is trained for task T_t and as $\mathbf{M}_i \in D_m$ is sampled from another distribution, $\widehat{\mathbf{M}}_i$ will be a meaningless reconstruction of $\mathbf{M}_i$. Alternatively, it can be said that the activations obtained on passing $\mathbf{M}_i \sim P(\mathbf{M})$ through the trained $f_t(\cdot, \theta_t)$ do not align with T_t. This can be observed from the increased MAE between $\mathbf{M}_i$ and $\widehat{\mathbf{M}}_i$ and from the t-SNE representations of the bottleneck layer (latent space) for several samples drawn from $P(\mathbf{M})$ (first two iterations of Fig. 8.2).

Let $\mathbf{F}_{\mathrm{r}i}$ be the output after a series of r *recursive iterations/passes* of $\mathbf{M}_i$ through the trained autoencoder $(f_t(\cdot, \theta_t))$, i.e.,

$$\mathbf{F}_{\mathrm{r}i} = f_t^r(f_t^{r-1}(\cdots f_t^1(\mathbf{M}_i, \theta_t) \cdots)). \quad (8.4)$$

Such recursive passing (output → input → output) gradually attunes the input M_i toward $F_{\mathrm{r}i}$ to produce activations that are accustomed to θ_t. This can be observed in Fig. 8.2, where the MAE reduces over recursive iterations and the latent space representations of samples drawn from $P(\mathbf{M})$ begin to overlap with those from $P(\mathbf{X})$. Let

$$\gamma_1 \leq |\mathbf{F}_{\mathrm{r}i} - \mathbf{F}_{(r-1)i}| \leq \gamma_2, \forall \mathbf{M}_i \in D_m, \mathbf{M}_i \sim P(\mathbf{M}), \quad (8.5)$$

where γ_1 and γ_2 are the lower and upper bounds of reconstruction error, respectively, for any $\mathbf{M}_i \in D_m$. The aim now is to have a collection set $F_t = \{\mathbf{F}_{\mathrm{r}i}\}_{i=1}^{N_f}$ for task T_t such that the following properties are satisfied:

Property 8.1. $\gamma_2 < \epsilon_1$ for $\forall \mathbf{M}_i \in D_m$.

168   *S. Gopalakrishnan et al.*

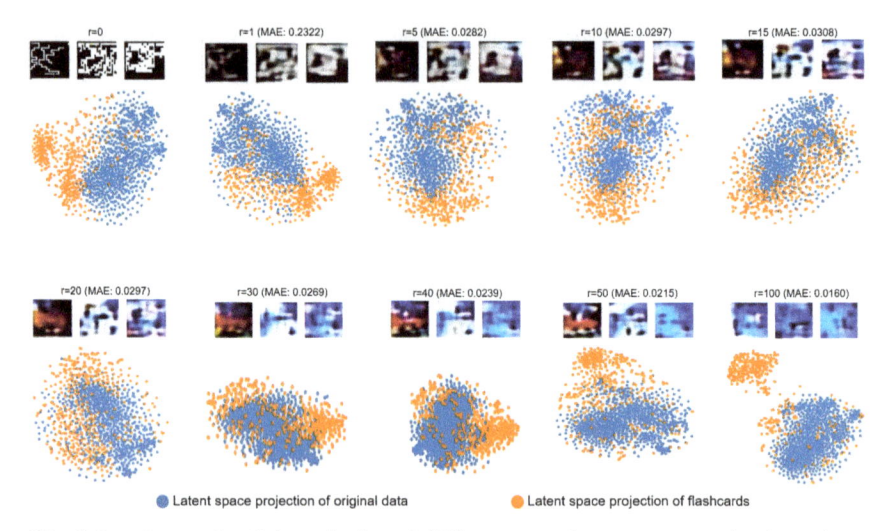

Fig. 8.2: A sample of three flashcard (F_t) constructions over recursive iterations, with Cifar10 as an example: Difference in MAE between successive iterations with iteration number and t-SNE of latent space in 2D. It can be observed that the initial (raw) input maze patterns adapt to the texture captured by the model as the number of recursive iterations increases (iteration $r = 10, 20$). Also, the corresponding latent space clusters get closer. However, on further repeated passes ($r \geq 50$), the reconstruction deteriorates/smooths out (as MAE keeps reducing), and the two clusters drift apart again (Property 8.1 is satisfied but Property 8.2 is not). Best viewed in color.

Property 8.2. Respective latent space distribution of $F_t \sim D_t$.

Such a set of constructions (F_t) obtained through recursive iterations of random inputs (drawn from $P(\mathbf{M})$) are defined as *flashcards*.

These flashcards are expected to capture the knowledge from the trained network and serve as a potential alternative to D_t. Figure 8.3 illustrates a few flashcards constructed with the same set of random images from autoencoders trained with different datasets. Although these flashcards are intended to capture the knowledge representations in $\boldsymbol{\theta}_t$ as a function of $\mathbf{M}_i \in D_m$, it can be observed from Fig. 8.3 that they do not bear direct shape similarities with their respective datasets. In fact, during the recursive process, the raw input image patterns are modified/transformed to textures that are suited to the trained autoencoder model (colored textures in flashcards are observed for Cifar10). This is in-line with the results

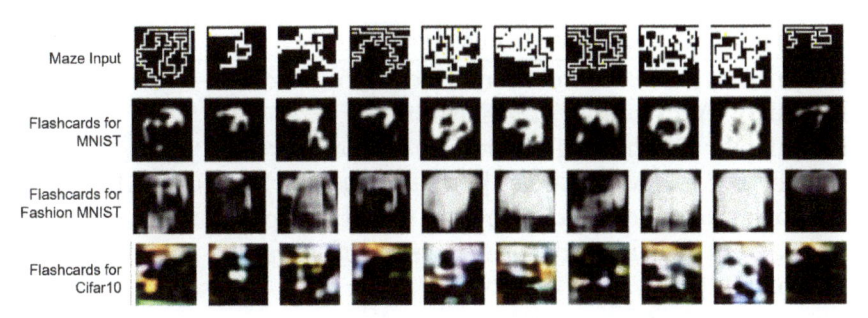

Fig. 8.3: Transformation from maze patterns (D_m) into flashcards (F_t) for tasks: MNIST, Fashion MNIST and Cifar10.

reported by Geirhos *et al.* (2019), which show that neural networks predominantly learn texture than shape by default. Now, since an autoencoder is trained to maximize $\log P(\mathbf{X}|\boldsymbol{\theta}_t)$, the following fact holds:

Fact 8.1. Let D_A and D_B be two i.i.d. datasets whose elements are drawn from $P(\mathbf{X})$. If there exists a trained autoencoder $f_A(\cdot, \boldsymbol{\theta}_A)$ to reconstruct D_A with error $[\beta_1, \beta_2]$, $\forall \mathbf{X} \in D_A$, then the autoencoder $f_A(\cdot, \boldsymbol{\theta}_A)$ can reconstruct D_B with the same error range.

The converse is not true because any image dataset D_B with images that have few pixels with non-zero values (other pixels as zeros) can still yield a smaller reconstruction error and not necessarily be drawn from $P(\mathbf{X})$ (satisfying Property 8.1). Hence, both Properties 8.1 and 8.2 for F_t need to be satisfied. As Properties 8.1 and 8.2 are dependent on the initial input distribution $P(\mathbf{M})$, it is important to have a suitable $P(\mathbf{M})$ to get F_t ($P(\mathbf{M}) = P(\mathbf{X})$ is one trivial and uninteresting solution). As finding a perfect $P(\mathbf{M})$ is an ideal research problem by itself, for all experiments in this chapter, random maze pattern images are used as a potential candidate D_m to get F_t. The above leads to the following hypothesis:

Hypothesis 8.1. Since the autoencoder trained on D_t also reconstructs elements from F_t (Property 8.1) and has matching latent space distribution (Property 8.2), flashcards (F_t) can be used as a set of pseudo-samples for D_t to learn a new autoencoder $f_t(\cdot, \boldsymbol{\theta}_{\text{new}})$, when trained from scratch using only F_t.

Algorithm 8.1 Flashcard construction.
Require: $f_t(\cdot, \boldsymbol{\theta}_t)$: Autoencoder model for task T_t
Require: N_f: Number of flashcards
Require: r: Number of recursive iterations through $f_t(\cdot, \boldsymbol{\theta}_t)$
 1: Let $D_m \leftarrow \{\mathbf{M}_1, \ldots \mathbf{M}_{N_f}\}$, where $\mathbf{M}_i \sim P(\mathbf{M}), \forall i$, $P(\mathbf{M})$ corresponds to the distribution of initial input patterns
 2: $F_t \leftarrow \{\}$, where F_t denotes the set of flashcards
 3: **for** $i \in \{1, \ldots, N_f\}$ **do**
 4: $\quad \mathbf{F}_{\mathrm{ri}} = f_t^r(f_t^{r-1}(\cdots f_t^1(\mathbf{M}_i, \boldsymbol{\theta}_t) \cdots))$
 5: $\quad F_t = F_t \cup \{\mathbf{F}_{\mathrm{ri}}\}$
 6: **end for**
 7: **return** F_t

8.2.1 Flashcard construction

Algorithm 8.1 provides the steps to construct flashcards using a trained autoencoder model. For flashcard construction, the following hyperparameters are useful: (i) the number of recursive iterations (r), (ii) the number of flashcards to construct (N_f) and (iii) the choice of initial input distribution $(P(\mathbf{M}))$. Figure 8.4 portrays an empirical analysis of selecting the optimal iteration, considering two datasets: SVHN and Cifar10. Figure 8.4 (Row 1) shows the Fréchet latent space distance (FLSD is inspired by FID (Heusel et al., 2017) and calculated as the Fréchet distance between latent space activations of flashcards versus original samples (smaller the better) from the eyes of the trained autoencoder) (orange) and MAE (blue), as a function of r. Initially, an increase in r leads to a decrease in FLSD up to certain iterations, which indicates that the flashcard distribution is getting closer to the learned distribution. As r increases further, though MAE keeps going down, FLSD increases because the propagation of reconstruction errors in the autoencoder causes a drift in the features.

8.2.1.1 Number of iterations

Based on empirical analysis, $10 \leq r \leq 20$ is a sensible choice that worked well across all conducted experiments. Reconstruction error (Fig. 8.4, Row 2) is also observed to be low in this iteration range.

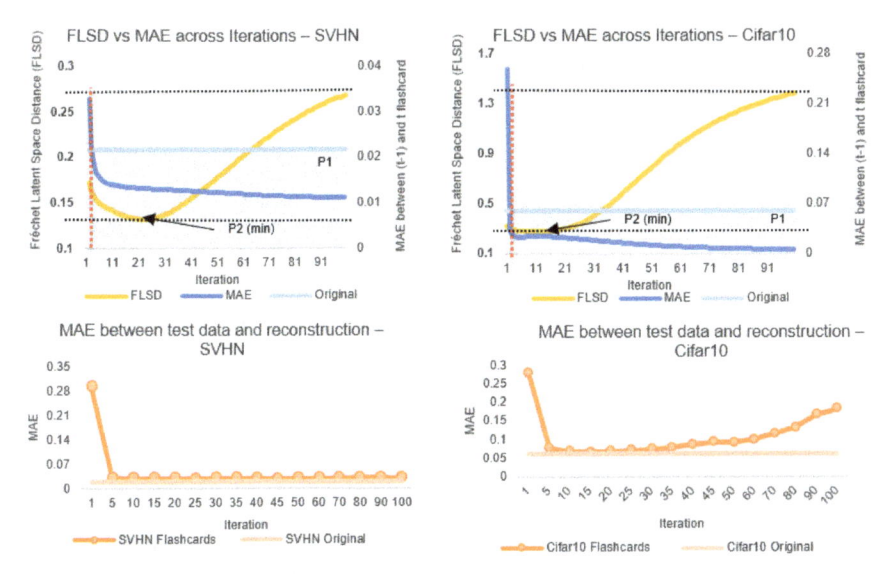

Fig. 8.4: Analysis on iterations r for flashcard construction, demonstrated using SVHN and Cifar10. Row 1 portrays metrics corresponding to the change on passing multiple random maze inputs in iterative fashion through an autoencoder. The orange curve depicts the trend of FLSD between intermediate input at iteration t and original data samples, and the dark-blue curve shows a constant decrease in reconstruction error between $t-1$ and t across different iterations for each dataset. The light-blue line serves as a reference MAE on the original data. Property 8.1 is satisfied when the blue curve goes below ϵ_1 (marked by the dotted red line), and Property 8.2 is satisfied at the first minima of FLSD, which occurs around $10 \leq r \leq 20$, irrespective of dataset. Row 2 compares the autoencoder error (MAE) between test data and its reconstruction when trained on flashcards constructed from different iterations. This again confirms the acceptable range of r and that flashcard construction is *not critically sensitive* to the choice of r within the range.

Table 8.1 provides the benchmark errors for using 50,000 flashcards obtained via different iterations when used for training a network from scratch. Results indicate that flashcards are not critically sensitive to r. In all experiments, r is set to 10. The general intuition of selecting r is to allow enough iterations to capture representations from the trained network and, at the same time, to keep in mind the irreducible error of the network that is added invariably with each pass.

Table 8.1: Benchmarking effect of iterations for reconstruction of different datasets, using 50,000 flashcards.

Dataset	Original	Flashcard 1 iteration	Flashcard 5 iterations	Flashcard 10 iterations	Flashcard 15 iterations
Cifar10	0.0564 ± 0.0014	0.1300 ± 0.0156	0.0708 ± 0.0027	0.0686 ± 0.0042	0.0740 ± 0.0097
MNIST	0.0184 ± 0.0014	0.0880 ± 0.0055	0.0675 ± 0.0007	0.0491 ± 0.0053	0.0417 ± 0.0027
Fashion MNIST	0.0259 ± 0.0003	0.1194 ± 0.0010	0.0437 ± 0.0003	0.0440 ± 0.0006	0.0465 ± 0.0019

8.2.1.2 *Number of flashcards*

Variations in flashcards can be obtained through the initial random sample chosen to be passed recursively through the autoencoder network. Thus, it is possible to generate countably infinite flashcards with subtle differences between them. An increase in the number of flashcards translates to performance improvement (up to a saturation threshold); examples using MNIST and Cifar10 are provided in Fig. 8.5. The trend followed by flashcards is similar to the improvement shown with increasing coreset/exemplars for training the network. Another example of improvement in network performance with increasing percentages of flashcards is shown for ImageNet 256 px later in Section 8.2.2.1.

8.2.1.3 *Flashcards initialization*

Finding the optimal $P(\mathbf{M})$ is a research question in itself. For experiments, the choice of $P(\mathbf{M})$ is maze-like random image patterns since there is a resemblance to edges and shapes. Gaussian random noise was also considered; however, there was a lack of diversity among the constructed flashcards. Using an additional dataset (such as next task images) did not help either. A study on the choice of initialization is presented in Table 8.2 and Fig. 8.6, using Sequence5, for continual reconstruction. Among the three choices, maze, Gaussian noise and next task data, only maze-based flashcards had diversity in terms of shape and texture and performed best due to their ability to capture activations at edges. Using Gaussian noise as initialization resulted in the repetition of the same patterns. Flashcards from new task

Flashcards for Knowledge Capture and Replay 173

Error reduction with increasing number of flash cards and core set samples

[Chart showing Mean Absolute Error (MAE) vs Number of Flash Cards / Coreset Samples]

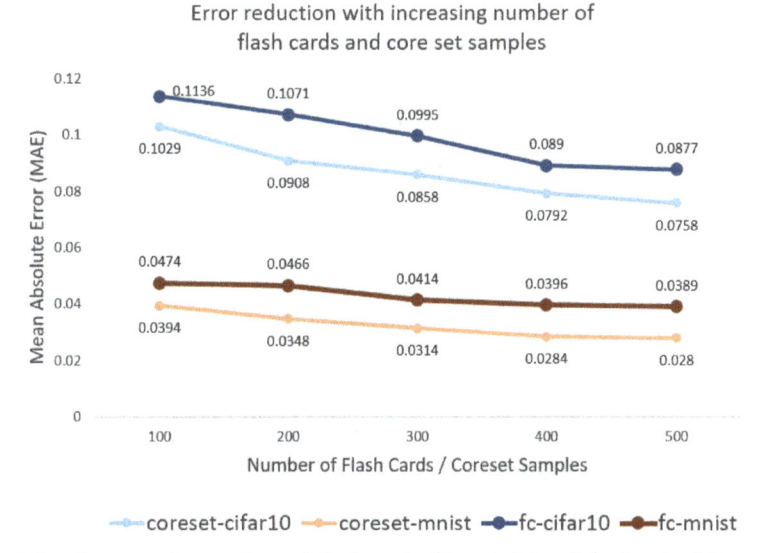

fc-cifar10: 0.1136, 0.1071, 0.0995, 0.089, 0.0877
coreset-cifar10: 0.1029, 0.0908, 0.0858, 0.0792, 0.0758
fc-mnist: 0.0474, 0.0466, 0.0414, 0.0396, 0.0389
coreset-mnist: 0.0394, 0.0348, 0.0314, 0.0284, 0.028

— coreset-cifar10   — coreset-mnist   — fc-cifar10   — fc-mnist

Fig. 8.5: Increase in number of flashcards (denoted by fc-) closely follows the improvement observed by training original (coreset) samples, as observed in (i) Cifar10 and (ii) MNIST.

Table 8.2: Maze patterns provide the best reconstruction MAE among the different initializations used for construction of flashcards. Experiments run on Sequence5 continual reconstruction.

Maze pattern	Gaussian noise	Next task data
0.0536 ± 0.002	0.0945 ± 0.003	0.3713 ± 0.275

initialization being dependent on the random subset images' coverage of old tasks' activations led to inconsistent results.

8.2.2 Flashcards for capturing representations

The feasibility of capturing single-task representations using flashcards is first demonstrated via verification of Hypothesis 8.1. The network's performance using flashcards as an alternative to the original dataset for training is first exhibited. Second, an attempt is made to see if flashcards can be used as a knowledge distillation mechanism. Third, the network trained using only flashcards is evaluated

174 S. Gopalakrishnan et al.

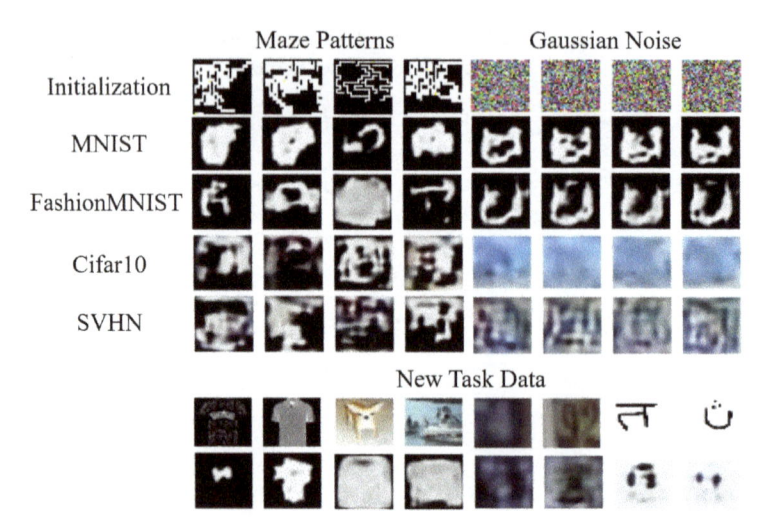

Fig. 8.6: Visual comparison of different initializations used for the construction of flashcards: Maze, Gaussian noise and next task. Flashcards constructed after training on a particular dataset subject to initialization are presented. Only maze-based flashcards have diversity in terms of shape and texture and perform best due to their ability to capture activations at edges. Gaussian noise results in repetitions. Flashcards from new task initialization are dependent on the random subset images' coverage of old tasks' activations and lead to inconsistent results (Table 8.2).

to see if the image reconstructions from this network capture enough discriminative (image) features to be used for classification.

8.2.2.1 Alternative to the original dataset

Consider three different datasets: MNIST, Fashion MNIST, and Cifar10. Let three different autoencoders (Network1), all with the same architecture, be trained separately for each of these three datasets using Eq. (8.1) as a loss function. Flashcards (F_t) constructed from each of these (Network1) autoencoders for the same set of random input maze pattern images (D_m) (a few samples shown in Fig. 8.3) are used, respectively, to train new autoencoder (Network2) models from scratch (with the same architecture). Results in Table 8.3 show reconstructions are very close to the autoencoders trained on originals.

Flashcards also work for large-scale resolution, as seen in Fig. 8.7. An autoencoder is trained on ImageNet 256 × 256 px resolution

Table 8.3: Reconstructions from two separate autoencoders, one trained on original data (Network1) and the other trained on flashcards (Network2). The metrics indicate that flashcards are sufficient alternatives for learning the original data. Weighted alpha (Martin and Mahoney, 2020) is based on HT-SR theory. Alpha values are closer for the trained networks, which indicate a similarity between the two network weights. The reported values are averaged over five runs.

Dataset	Network 1's MAE	Network 2's MAE	Alpha original	Alpha flashcards	Alpha untrained
MNIST	0.0184	0.0491	1.9185	2.0531	1.4343
Fashion	0.0259	0.0440	2.0044	2.1158	1.4343
Cifar10	0.0564	0.0686	2.0266	2.1531	1.4343

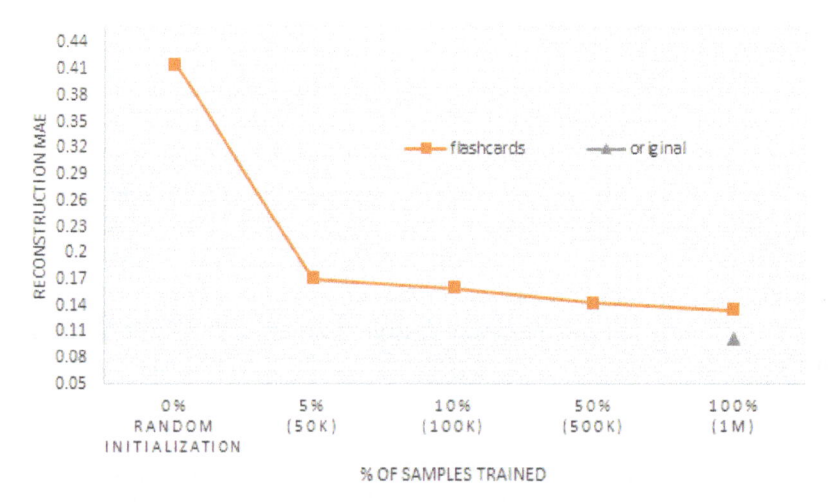

Fig. 8.7: Reconstruction error of two networks, one trained on original ImageNet 256 × 256 resolution 1M images (gray), and the second trained only on flashcards 256 × 256 resolution (orange). The x-axis denotes the percentage of samples added, and as more flashcards are introduced, the error in the second network approaches closer to that of the first.

(1M images) and has a reconstruction error of 0.10. A separate autoencoder is trained using only flashcards of varying percentages. The error when the network is randomly initialized without any data is high (0.42); however, with an increase in the number of flashcards, it drops from 0.17 to 0.13, matching closely with the original benchmark.

Results from Table 8.3 and Fig. 8.7 confirm the hypothesis that flashcards indeed capture network parameters as a function of D_m and therefore can be used as pseudo-samples/training data for both small- and large-scale resolutions (as network parameters involved in flashcard construction are initially learned by training with the original dataset D_t).

8.2.2.2 Data-free knowledge distillation

Consider training an autoencoder teacher network AE1 with data D_{T_1}. Once trained, D_{T_1} may become unavailable due to certain constraints, such as confidentiality/privacy or storage requirements. Regular knowledge distillation (Gou et al., 2020) is no longer possible because of the absence of original data while distilling for the student network. If in the future there is a newer and better architecture AE2, migration is possible by training on flashcards constructed from AE1 to AE2 (data-free knowledge distillation). Table 8.4 shows feasibility by distilling knowledge from teacher AE1 to student AE2. The student AE2 is selected as one smaller and one larger modified architecture to the teacher. One can observe that distillation and transfer are still achievable without original data, and an improvement in performance when selecting larger architectures indicates that flashcards enable migration to better models in the future.

Table 8.4: Building autoencoder AE2 (smaller and larger arch.) using flashcards obtained from AE1 trained on the Cifar10 dataset. The column "Test original" shows test MAE when these architectures are trained using the original Cifar10 dataset. The column "Test AE2" refers to test MAE on Cifar10 when flashcards obtained from AE1 are used to train AE2 for reconstruction. Better performance on both the AE2-Smaller and AE2-Larger architectures showcases transfer using flashcards.

Train arch. type	Params.	Test original	Test AE2
AE1 - Original	94,243	0.0640	—
AE2 - Smaller arch.	24,083	0.0787	0.0963
AE2 - Larger arch.	372,803	0.0512	0.0570

8.2.2.3 Evaluating flashcards derived autoencoder: Are the reconstructions good enough to be used for classification?

Let AE1 be trained on the original images D_{T_1} and AE-Flash1 be trained using the flashcards D_{f_1} from AE1. Let the respective reconstructions (after training) be $\widehat{D_{T_1}}$ and $\widehat{D_{F_1}}$. Next, train two classifiers (VGG16), Classifier 1 and Classifier 2, using $\widehat{D_{T_1}}$ and $\widehat{D_{F_1}}$, respectively, and compare their performance on an independent test set of original images. The results tabulated in Table 8.5 show that flashcard-trained autoencoder networks are capable of providing reconstructions that perform reasonably well as inputs for classification.

8.2.3 Flashcards for replay in continual learning

Consider a sequence of T tasks $\{T_1, \ldots, T_t, \ldots, T_T\}$. In the CL scenario, the network for task T_{t+1} is required to be trained on top of previously learned tasks $T_1, \ldots, T_t$. In other words, $\boldsymbol{\theta}_{t+1}$ is adapted from the previously trained network parameters $\boldsymbol{\theta}_t$. Training for task T_{t+1} may result in the network forgetting the representations learned until the previous task T_t. Unlike other CL-based approaches, which aim to preserve $\boldsymbol{\theta}_t$ through regularization, data replay (either episodic or using external generative networks), architectural strategies, or their combinations, we use the flashcards constructed on $\boldsymbol{\theta}_t$ along with data for task T_{t+1}, while training for task T_{t+1}. Flashcards are required to be constructed only at the end of task T_t, *irrespective* of the number of preceding tasks, so that knowledge representations for tasks $\{T_1, \ldots, T_t\}$ can be captured and trained with the next

Table 8.5: Building a classifier using the reconstructions from flashcard-trained network. The reported accuracies are averaged over five runs. Cifar10 accuracy is lower due to a relatively higher autoencoder reconstruction error.

Dataset	Original	Classifier 1	Classifier 2
MNIST	0.9940	0.9921	0.9869
Fashion MNIST	0.9215	0.9126	0.8978
Cifar10	0.8003	0.6295	0.5659

Algorithm 8.2 Flashcard for replay in continual learning.

Require: $f_t(\cdot, \boldsymbol{\theta}_t)$: Autoencoder model trained till task T_t
Require: N_f: Number of flashcards
Require: r: Number of iterations through $f_t(\cdot, \boldsymbol{\theta}_t)$
Require: D_{t+1}: Data for task T_{t+1}
Require: λ: Scaling parameter, by default, set to 1 (in all Exp.)
1: $F_t \leftarrow$ **Algorithm 8.1** $(f_t(\cdot, \boldsymbol{\theta}_t), N_f, r)$, where F_t is the set of flashcards for task T_t
2: Train the autoencoder model $f_{t+1}(., \boldsymbol{\theta}_{t+1})$ on flashcards F_t and data D_{t+1}
3: Initialize $\boldsymbol{\theta}_{t+1} \leftarrow \boldsymbol{\theta}_t$
4: Optimize $\boldsymbol{\theta}_{t+1}$ using given loss

$$\boldsymbol{\theta}^*_{t+1} \leftarrow \arg\min_{\boldsymbol{\theta}_{t+1}} \left(\frac{1}{|D_{t+1}|} \sum_{n=1}^{|D_{t+1}|} |D_{t+1}^{(n)} - f_{t+1}(D_{t+1}^{(n)}, \boldsymbol{\theta}_{t+1})| \right.$$
$$\left. + \frac{\lambda}{N_f} \sum_{n=1}^{N_f} |F_t^{(n)} - f_{t+1}(F_t^{(n)}, \boldsymbol{\theta}_{t+1})| \right)$$

where $D_{t+1}^{(n)}$ and $F_t^{(n)}$ are the nth data sample and flashcard sample, respectively
5: **return** $f_{t+1}(., \boldsymbol{\theta}^*_{t+1})$

---

task T_{t+1}. Thus, the proposed method avoids storing flashcards for each successive task, thereby significantly reducing the memory overhead while ensuring robust performance (experiments in Section 8.4). Algorithm 8.2 provides the steps for using flashcards as a replay strategy for CL. Flashcards are constructed on the fly before training a new task (irrespective of the number of previous tasks).

8.3 Comparison with Other Continual Learning Algorithms

In general, CL algorithms are based on architectural strategies, regularization, memory replay, and their combinations (Parisi et al., 2019). Owing to their intrinsic nature (as discussed in Sections 8.2.2 and 8.2.3), flashcards implicitly exhibit the characteristics of both regularization (refer to Algorithm 8.2 point 4)

Table 8.6: Comparison between different continual learning replay strategies.

Trait	Flashcards for replay	Episodic memory replay	Generative (data) replay
Examples			
Single-task scenario			
Alternative to original data?	Yes	No, subset of original	Yes
Distinction	Postprocess knowledge capture from trained network	Stores a subset of the original data in memory	Network is trained to generate samples per task
Visually similar to dataset?	No	Yes	Depends on data complexity
Performance on task	Close to original network	Very close to original network	Close to original network
Continual learning scenario			
Store sample across tasks?	No	Yes, need to store per task	Depends on nw. complexity
Nw. used to obtain alt. data	Simple autoencoder	N/A	VAE/GAN, etc.
Store between tasks?	No, just-in-time creation	Yes, either memory or disk	Depends on support for CL
Constructed on the fly?	Yes, just before starting next task training	No, store original samples in memory	On the fly (if CL supported) or each task's exclusive nw.
Independent of no. of tasks?	Yes, need current snapshot only	No, need to store per task	Depends on support for CL
Scaling up tasks (complexity)	Results shows feasibility	Straightforward: linear storage	Challenging with more tasks

and replay and will therefore be compared with the respective strategies. Regularization strategies, such as those used by Kirkpatrick *et al.* (2017) and Zenke *et al.* (2017), learn new tasks while imposing constraints on the network parameters to avoid deviating too much from those learned from the previous tasks. Li and Hoiem (2018) use current task samples to regularize the soft labels of past tasks, thus avoiding explicit data storage. However, it is important to note that these regularization methods perform well in homogeneous (sharing similar characteristics) task environments, where it is possible to find mutually suboptimal points in the solution space. However, such methods suffer when subsequent tasks come from different domains/heterogeneous datasets (Li and Hoiem, 2018).

In a rehearsal mechanism, a subset of samples from previous tasks, referred to as coreset samples/exemplars, serve as memory replay (Castro *et al.*, 2018; Chaudhry *et al.*, 2018b; Lopez-Paz and Ranzato, 2017). However, they suffer from excessive storage requirements that grow with the number of tasks. On the other hand, generative replay approaches (Aljundi *et al.*, 2019a; Li *et al.*, 2020; van de Ven *et al.*, 2020; Rostami *et al.*, 2020) face challenges in scalability for complex datasets with many tasks since they involve heavy training and computation associated with an auxiliary network to generate visually meaningful images for each task.

The approaches discussed above are either suitable only for homogeneous data, are memory intensive as they involve *preserving* samples, are computationally intensive in "generating" samples, or require task identifiers. The proposed flashcards are constructed on the fly, require only a simple autoencoder, are independent of the number of tasks and are helpful *both* as regularization and replay mechanisms. Flashcards can be used as *pseudo-samples* constructed with low computational and memory expenses. Furthermore, flashcards can scale across tasks without a considerable drop in performance. Table 8.6 lists an overall comparison of various traits between flashcards, episodic, and generative replay.

8.4 Experiments in Continual Learning

This section provides experimental validation of using flashcards in a CL scenario as a replay method for different applications, such as: (i) continual reconstruction (Section 8.4.1), (ii) continual denoising

(Section 8.4.2), new task incremental classification (Section 8.4.3) and single-task new instance learning (Section 8.4.4). All experiments are compared with the upper bound (joint training, JT), the lower bound (sequential fine-tuning, SFT), and standard methods for regularization, generative and episodic replay as baselines. Results indicate that using flashcards indeed helps under heterogeneous conditions spanning different domains and where the task identifier is not available.

Datasets: For continual reconstruction and denoising, five heterogeneous public datasets, referred to as Sequence5, were employed. The datasets used in Sequence5 are in the following order: MNIST, Fashion MNIST, Cifar10, SVHN and Omniglot. Each dataset was considered a task, with class labels omitted. MNIST, Fashion MNIST and Omniglot were resized to $32 \times 32 \times 3$ (bilinear rescaled and channel copied twice) to maintain the same scale as the other two datasets. Continual reconstruction is also compared on UC Merced land use (Yang and Newsam, 2010), a real-world large-scale remote sensing dataset with 256×256 resolution and 21 classes. For task-incremental classification, Sequence3 comprising the order of Cifar10, MNIST and Fashion MNIST was employed. For a new instance of learning, Cifar10, was selected with different brightness and saturation variations.

Architecture and Training Details: For reconstruction and denoising, flashcards are constructed using an autoencoder architecture with four layers of down/upsampling and 64 convolutional filters per layer. The bottleneck dimension of 256 offers 12× reduction in image space, serving to demonstrate the effectiveness of forgetting. The architecture is a simplified variant of the VGG, adapted with fewer layers and filters for proper reconstruction. All hidden layers employ *tanh* activation.

Several autoencoder architectures were trained on the Cifar10 dataset to compare the performance of flashcards for reconstruction. Table 8.7 provides details about various model architectures and the corresponding test MAE on training using the original dataset (Original MAE) and using flashcards generated from the trained AE (Flashcards MAE), respectively. The architecture *Blk_4_fil_64* was chosen for experiments in reconstruction and denoising tasks. *Blk_4_fil_64* architecture obtains 0.0512 Original MAE and

Table 8.7: Architecture selection for autoencoder (AE). Several AE architectures were trained on the Cifar10 dataset in order to compare the performance of flashcards for reconstruction. Various details about the architecture, such as Model Params. (number of trainable weights and biases), Latent Space (size of latent space/bottleneck layer and its reduction rate versus image space), Num. Blocks (number of convolution + pooling blocks in encoder), and Num. Filters (number of filters in convolution layers) are also provided. Original MAE is the Cifar10 test MAE on AE trained using the Cifar10 train dataset. Flashcards MAE is the Cifar10 test MAE on the AE trained using the flashcards obtained from the given AE. The reported standard deviation for the scores was obtained over five experimental runs.

Arch. type	Model params.	Latent space	Num. blocks	Num. filters	Original MAE	Flashcards MAE
Blk_4_fil_16	24,083	64 (48 × reduction)	4	16	0.0787 ± 0.0002	0.0963 ± 0.0004
Blk_4_fil_32	94,243	128 (24 × reduction)	4	32	0.0640 ± 0.0002	0.0725 ± 0.0009
Blk_4_fil_64	372,803	**256 (12 × reduction)**	4	64	0.0512 ± 0.0004	0.0570 ± 0.0006
Blk_4_fil_128	1,482,883	512 (6 × reduction)	4	128	0.2062 ± 0.0000	0.2445 ± 0.0476
Blk_3_fil_64	298,947	1024 (3 × reduction)	3	64	0.0437 ± 0.0003	0.0599 ± 0.0067
Blk_2_fil_32	57,251	2048 (1.5 × reduction)	2	32	0.0358 ± 0.0008	0.0389 ± 0.0015

0.0570 Flashcards MAE. *Blk_3_fil_64* and *Blk_2_fil_32* achieve better Original MAE/Flashcards MAE than *Blk_4_fil_64*; however, both these architectures use a higher latent space size (1024 and 2048). The improvement in reconstruction may be attributed to the higher latent space dimension, which might not encode useful information but act as a copy function.

Training Details: For reconstruction and denoising, the network is trained for 100 epochs per task and optimized using Adam with a learning rate of 0.001. Each minibatch update is based on an equal number of flashcards and the current task samples. Of the training data, 10% is allocated for validation. Early stopping is employed if there is no improvement for over 20 epochs. 5K flashcards with $\lambda = 1$ (refer to Algorithm 8.2) were employed for replay.

Redefining Metrics: For reconstruction and denoising, individual task MAE is measured before and after observing the data for the given task, as is the ability to transfer knowledge from one task to another by measuring backward transfer (BWT). A similar definition is followed for average MAE and BWT, as described by Lopez-Paz and Ranzato (2017). It must be noted that Lopez-Paz and Ranzato (2017) describe these definitions in a supervised setting. With test sets for each of the T tasks, evaluate the model obtained after training task T_t on all T tasks. This gives the matrix $M \in R^{T \times T}$, where $M_{i,j}$ represents the test MAE for the model on task T_j after observing the data on task T_i. Then, redefine the following metrics:

$$\text{Average MAE} = \frac{1}{T} \sum_{i=1}^{T} M_{T,i}, \tag{8.6}$$

$$\text{BWT} = \frac{1}{T-1} \sum_{i=1}^{T-1} (M_{i,i} - M_{T,i}). \tag{8.7}$$

Lower values for average MAE and higher values for BWT are better. For task incremental classification and single-task, new-instance learning, the accuracy averaged over all past and current tasks is considered.

Table 8.8: Continual reconstruction for homogeneous tasks (each task T_k specifies up to k classes from the same dataset that have been observed) shows minimal/no forgetting, with examples showcasing reconstruction error for Cifar10 and Caltech101. On the contrary, there is a strong indication that knowledge from the past improves learning newer tasks (forward transfer).

Method	Cifar10 T_1	T_5	T_{10}	Caltech101 T_1	T_{50}	T_{101}
JT	—	—	0.0596	—	—	0.0831
SFT	0.0872	0.0748	0.0719	0.1088	0.1002	0.0906

8.4.1 *Heterogeneous continual reconstruction*

Since continual reconstruction for homogeneous cases is trivial and does not require extra effort (as observed from Table 8.8), the focus is more on the challenging heterogeneous scenario where forgetting is inevitable. Comparison with baseline methods is summarized in the captions of Table 8.12. Flashcards compared against different continual learning-based regularization and replay techniques outperform most baselines using fewer network capacity. It performs on par with episodic replay methods without occupying external memory for past samples.

8.4.1.1 *Sequence3 continual reconstruction*

Starting with a simpler example, an experiment comprising different permutations of MNIST, Fashion MNIST and Cifar10 (collectively called Sequence3) substantiates the effect of flashcards based on the order of the tasks seen. This allows us to study with more granularity how the order of the previous task affects the learning of the current task. Tables 8.9–8.11 compare the mitigation of forgetting with reference to the upper/lower bounds, coreset sampling and flashcards, respectively.

8.4.1.2 *Sequence5 continual reconstruction*

The experiment is further made challenging by considering two more tasks: SVHN and Omniglot, with the five tasks collectively termed

Table 8.9: Sequence3 order: MNIST, Fashion MNIST and Cifar10 as three tasks. Tasks are added incrementally, and MAE is computed on each dataset after the current task is completed.

Method	Task	MNIST	Fashion MNIST	Cifar10	Avg. MAE
Joint training	—	0.0141	0.0256	0.0629	0.0342
Coreset sampling 5000	1	0.0190	—	—	0.0190
	2	0.0245	0.0388	—	0.0316
	3	0.0249	0.0395	0.0666	0.0436
Lower bound	1	0.0190	—	—	0.0190
	2	0.0268	0.0268	—	0.0268
	3	0.0467	0.0469	0.0512	0.0482
Flashcards 5000	1	0.0190	—	—	0.0190
	2	0.0243	0.0310	—	0.0276
	3	0.0282	0.0366	0.0579	**0.0409**

Table 8.10: Sequence3 order: Fashion MNIST, Cifar10 and MNIST as three tasks.

Method	Task	Fashion MNIST	Cifar10	MNIST	Avg. MAE
Joint training	—	0.0141	0.0256	0.0629	0.0342
Coreset sampling 5000	1	0.0324	—	—	0.0324
	2	0.0344	0.0589	—	0.0466
	3	0.0386	0.0661	0.0200	0.0415
Lower Bound	1	0.0324	—	—	0.0324
	2	0.0548	0.0564	—	0.0556
	3	0.0816	0.2996	0.0140	0.1317
Flashcards 5000	1	0.0324	—	—	0.0324
	2	0.0336	0.0520	—	0.0428
	3	0.0352	0.0637	0.0156	**0.0381**

as Sequence5. The order of the tasks is as follows: MNIST, Fashion MNIST, Cifar10, SVHN and Omniglot. A visualization of the reconstructed samples at the end of the last task is provided in Fig. 8.8. Sequential fine-tuning (SFT), being a naive approach, suffers the most from forgetting, and most reconstructions are empty. Further, the weight initialization of the network parameters at the start of task 5 prevents it from learning the current Omniglot task itself. Continual VAE has partial reconstruction and also suffers from the

Table 8.11: Sequence3 order: Cifar10, MNIST and Fashion MNIST as three tasks.

Method	Task	Cifar10	MNIST	Fashion MNIST	Avg. MAE
Joint training	—	0.0141	0.0256	0.0629	0.0342
	1	0.0515	—	—	0.0515
Coreset sampling 5000	2	0.0639	0.0220	—	0.0429
	3	0.0654	0.0229	0.0336	**0.0406**
	1	0.0515	—	—	0.0515
Lower bound	2	0.2602	0.0142	—	0.1372
	3	0.1233	0.0465	0.0371	0.0689
	1	0.0515	—	—	0.0515
Flashcards 5000	2	0.0625	0.0181	—	0.0403
	3	0.0664	0.0261	0.0308	0.0411

loss of certain textures and colors. Using flashcards, past and current task samples are properly remembered.

Visuals are also provided for each task in Sequence5 showing how each method handles forgetting. Figure 8.9 shows sequential fine-tuning (SFT) is the naive approach and suffers the most. It can be observed that reconstructions are empty; this is because of the network parameters at the start of task 5, which prevent it from learning the current Omniglot task itself. Figure 8.10 shows the effect of replay with 500 real samples (coreset). Five hundred samples were chosen because their memory matches the AE network parameters of 1.5 MB. From the experimental results, it is observed that 500 samples are not sufficient to beat flashcards. Figure 8.11 is based on VAE trained in CL fashion, maintaining the same mean and standard deviation across tasks. It is not sufficient to mitigate forgetting. Figure 8.12 uses an autoencoder for reconstruction, supplemented by an external VAE for generative replay. Though results are competitive with flashcards, there is still forgetting in the previous tasks of MNIST and Fashion MNIST. Figure 8.13 presents results when using flashcards, where the past and current task samples are well remembered. Figure 8.14, with individual graphs for different methods, shows the variation of test MAE on the current task dataset after observing the data for the sequence of tasks.

UC Merced: An example with a remote sensing dataset: Continual reconstruction results over 21 tasks on UC Merced land

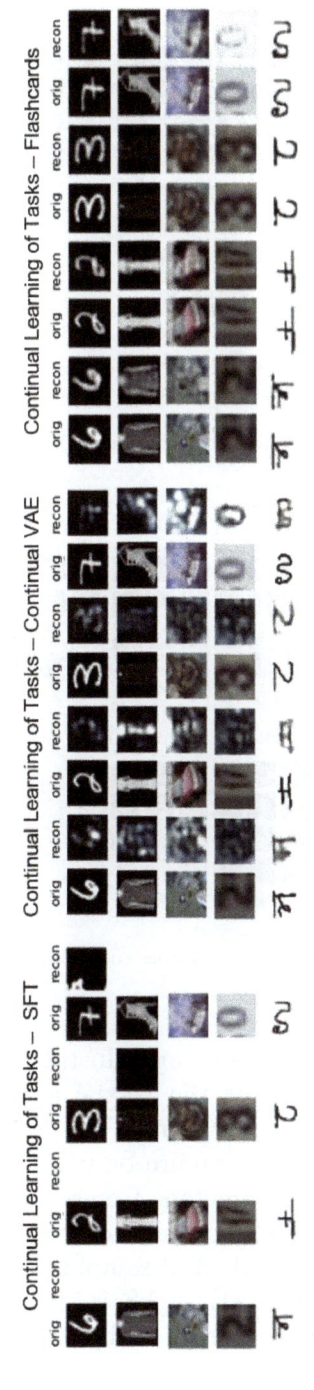

Fig. 8.8: Comparison between different methods for continual reconstruction using Sequence5. The figure shows the impact of forgetting after the end of the *last* task, i.e., Omniglot (all intermediate results are in Supplementary). Each row presents a dataset/task. SFT (left) is not able to reconstruct, even when presented with Omniglot, because of (i) catastrophic forgetting and (ii) unsuitable transfer learned weight initialization. Continual VAE (center) fares better but has artifacts and color loss in the reconstructions. Flashcards (right) remember all previous tasks.

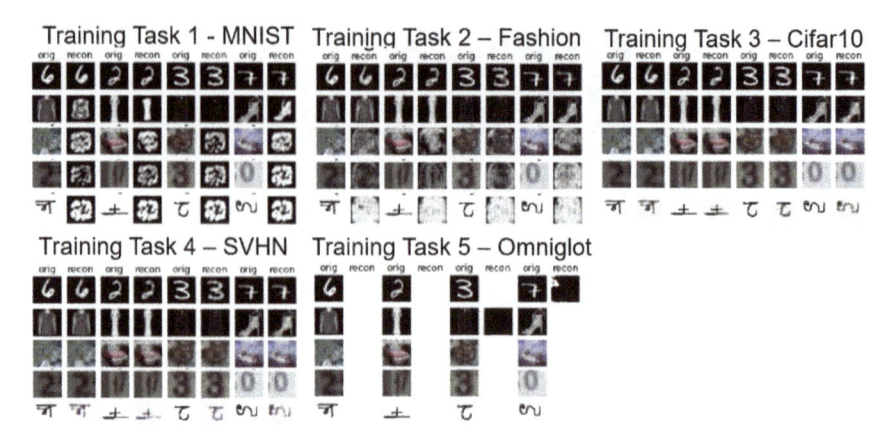

Fig. 8.9: Continual reconstruction on naive/sequential fine-tuning (SFT).

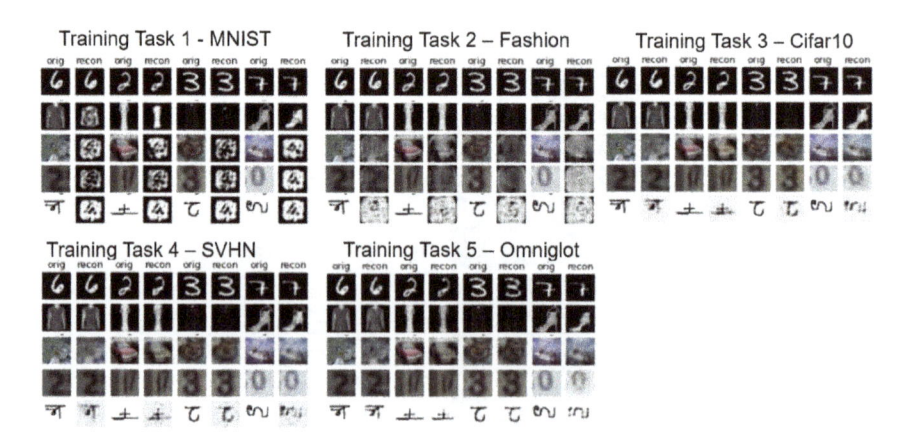

Fig. 8.10: Continual reconstruction using episodic memory: Coreset 500.

use remote sensing dataset (256 px) are illustrated in Fig. 8.15. A comparison is made per task (on the x-axis) against SFT, coreset (maximum 100 samples in buffer), and VAE + AE, with JT as a reference line. All autoencoders require only 2.9 MB (for 256 × 256 dim) in terms of network parameters. Coreset requires an extra 75 MB to store 100 images. VAE occupies an extra 4.7 MB. Both flashcards and VAE + AE generate 100 samples for the next task and have similar performance. VAE + AE is slightly better than flashcards at the expense of more storage. Other methods experience

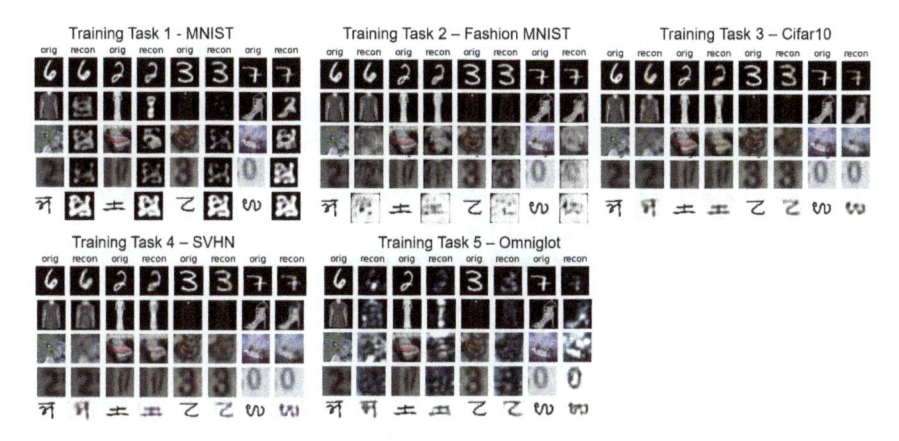

Fig. 8.11: Continual reconstruction using VAE trained exclusively for CL.

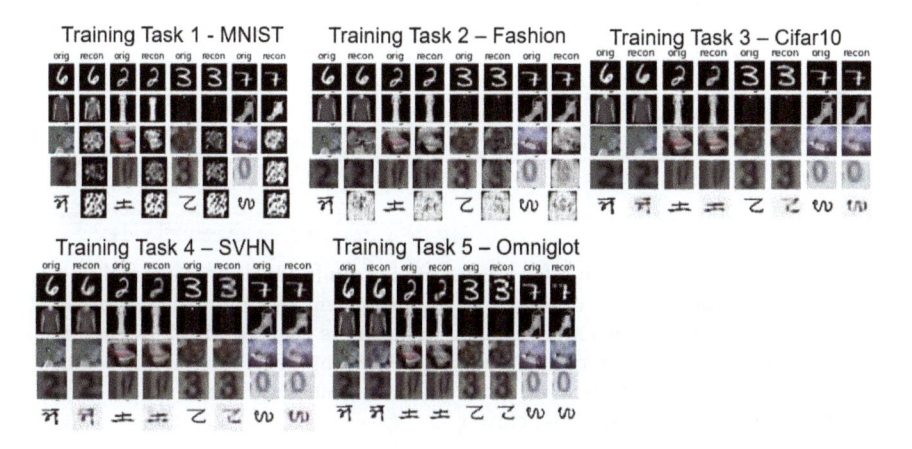

Fig. 8.12: Continual reconstruction using AE + VAE as generative replay.

a spike in transition to the next task, whereas performance is stable when using flashcards.

8.4.2 *Heterogeneous continual denoising*

A more challenging extension to reconstruction is denoising in a continual learning scenario. Here, the objective of the network is to learn, remember and denoise the samples simultaneously. The noise imposed on images is sampled from a standard normal distribution factored by a small parameter in the Sequence5 datasets. Results

190   S. Gopalakrishnan et al.

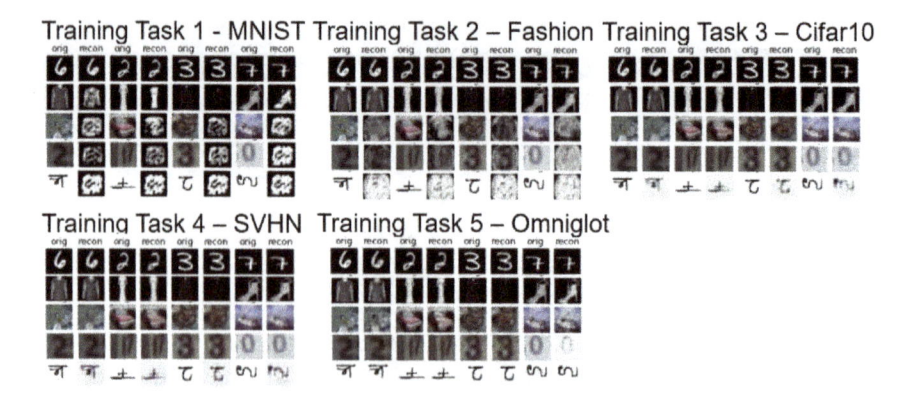

Fig. 8.13: Continual reconstruction using flashcards.

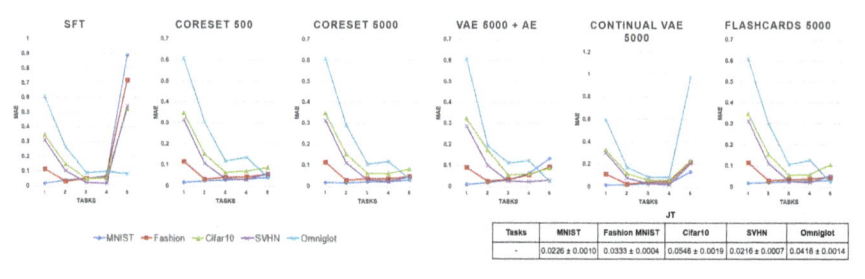

Fig. 8.14: Continual learning for reconstruction. Individual graphs for different methods show the variation of test MAE on the current task dataset after observing the data for a sequence of tasks. The table shows the test MAE for the joint training (JT) method. The reported values in the graph and table were obtained over five experimental runs. The standard deviation is quite small and is not displayed on the graphs to avoid clutter.

reported for Denoise MAE and BWT under columns 7 and 8 in Table 8.12 are obtained by adding a noise factor of 0.1 to the original images. Flashcard replay performs better than baseline approaches, with lower error compared to regularization and generative replay methods, even outperforms the episodic replay method: Coreset 5K in terms of MAE.

To study how robust the method works, the noise factor was steadily increased to check for the value where reconstruction fails completely. Figure 8.16 shows the impact of reconstruction using flashcards for different noise level settings. As more noise is added, it becomes visually difficult to make out the underlying image.

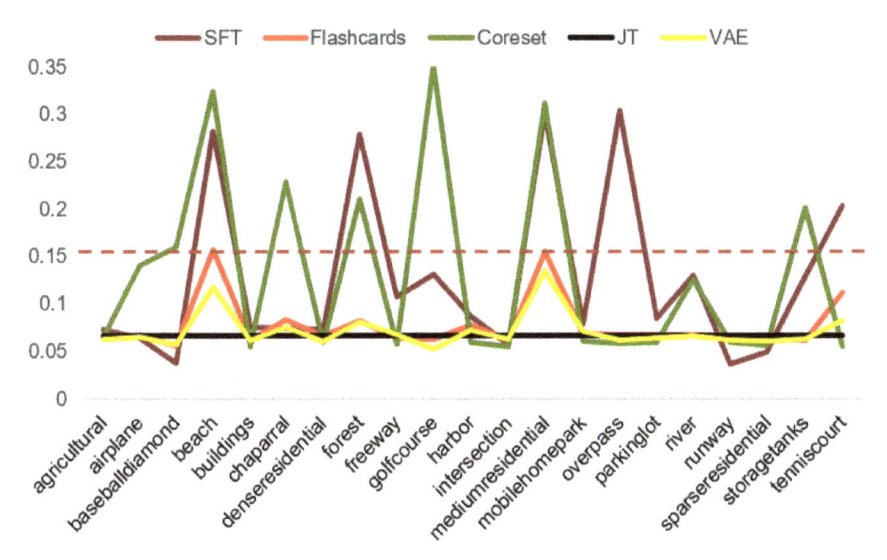

Fig. 8.15: Continual reconstruction on the UC Merced land use 256 px dataset. The line plot shows reconstruction error averaged from the first to the current task, as new tasks are introduced. SFT and coreset oscillate occasionally as new tasks are encountered, e.g., beach (blue texture) after agriculture and baseball ground (green), forest after series of buildings (gray), golf course after freeway, overpass after park, etc. Despite minimal capacity among other methods, flashcards have a relatively stable transition (dotted horizontal line shows peak error) and low error for all tasks.

At a factor of 0.3, it is observed that the network is trying to retain a partial outer boundary but has forgotten the denoising ability when seeing the last task, Omniglot.

8.4.3 *Task (agnostic) incremental classification*

Task-incremental learning (Task-IL) is a continual learning setting where new tasks, each comprising multiple classes (≥ 2), are learned in an incremental fashion. In traditional Task-IL, the network encoder is shared, whereas there exist separate task-exclusive multi-head outputs. Generally, classification is performed using a specific multi-head with the help of a task identifier. In a task-agnostic setting, the decision is made either by a smart selection of the multi-head (replacing the oracle/task identifier) or by taking the most confident prediction among all heads.

Table 8.12: Continual learning for Sequence5 reconstruction and denoising, compared with different methods in terms of error observed in MAE and BWT on the test set. Additionally compared are the network capacity (N/w in MB) and external memory (Mem in MB) required by each method.

Method	Type	N/w	Mem	Recon. MAE ↓	Recon. BWT ↑	Denoise MAE ↓	Denoise BWT ↑
JT	–	1.5	798.7	0.0348 ± 0.000	–	0.0552 ± 0.000	–
SFT	–	1.5	–	0.5518 ± 0.001	−0.6385 ± 0.002	0.5477 ± 0.002	−0.6275 ± 0.002
Coreset 500	ER	1.5	1.5	0.0580 ± 0.005	−0.0246 ± 0.006	0.0748 ± 0.004	−0.0219 ± 0.005
Coreset 5000	ER	1.5	15.3	0.0495 ± 0.002	−0.0185 ± 0.002	0.0595 ± 0.002	−0.0172 ± 0.002
LwF	Reg + R	1.5	–	0.5495 ± 0.002	−0.5241 ± 0.002	0.4370 ± 0.002	−0.3338 ± 0.002
VAE 5000 + AE	GR	2.9	–	0.0751 ± 0.004	−0.0575 ± 0.006	0.0711 ± 0.005	−0.0438 ± 0.005
CL VAE 5000	GR	1.4	–	0.3524 ± 0.041	−0.1658 ± 0.038	0.5460 ± 0.042	−0.2839 ± 0.033
Flashcards 5000	FR	**1.5**	–	**0.0536 ± 0.002**	**−0.0277 ± 0.002**	**0.0588 ± 0.002**	**−0.0262 ± 0.002**

Notes: Flashcards perform better than generative method-based replay and are on par with the error of using episodic memory-based coreset replay. Further, the memory requirement is very low, and an external memory buffer to store past tasks is not required. Lower avg. MAE (↓) and higher BWT (↑) are better. Reg: regularization, Reg + R: regularization and replay, ER: episodic replay, GR: generative replay, FR: flashcard replay. Reported scores are averaged over three independent runs.

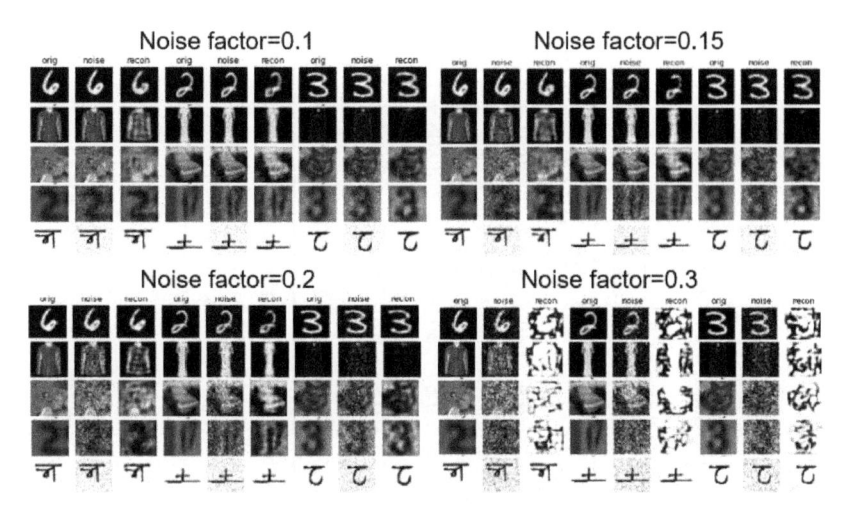

Fig. 8.16: Continual denoising scenario. Shown in the figure is the effect of noise applied and the reconstruction of Sequence5 using flashcards.

The Task-IL experiment explained here focuses on a heterogeneous scenario where each task has a different data distribution. A comparison is made for both cases of with and without task identifier. The same setup as provided by van de Ven et al. (2020) was followed. For the architecture of the autoencoder, the encoder weights are pre-initialized with classifier encoder weights, a decoder is attached after latent space, and the autoencoder is trained by passing current task samples along with flashcards up to the current task. According to Geirhos et al. (2019), "neural networks learn predominantly texture than shape by default," which is especially true in a heterogeneous setting where there is minimal/no overlap of properties among the tasks. Autoencoders learn to capture the texture behind a task as an essential feature for reconstruction; hence, using an autoencoder in the loop helps to mitigate forgetting of texture. Using a classification-trained encoder and pretrained weights as initialization for the autoencoder serves as a good starting point to retain classification features, and at the same time, the weight update is within range to relearn the relevant texture of the task on further training of a few epochs. For the next task, the classifier uses new task samples along with constructed flashcards of the previous task. Soft labels for flashcards are obtained by passing them through the classifier before the start of the next task. A mini-batch size of 256

Table 8.13: Comparison of different methods for task-incremental learning using Sequence3 tasks (Cifar10–MNIST–Fashion MNIST). The reported values are accuracies computed at the end of task T.

Method	Type	With task identifier			Without task identifier		
		T1	T1→T2	T1→T2→T3	T1	T1→T2	T1→T2→T3
SFT	–	61.55 ± 0.8	60.20 ± 0.1	43.55 ± 0.4	61.55 ± 0.8	49.41 ± 0.1	30.21 ± 0.2
Coreset 100%	ER	61.55 ± 0.8	79.26 ± 0.6	83.13 ± 0.8	61.55 ± 0.8	81.15 ± 1.1	84.38 ± 1.0
EWC	Reg	61.55 ± 0.8	66.50 ± 2.9	72.75 ± 1.9	61.55 ± 0.8	50.27 ± 3.2	28.18 ± 2.2
SI	Reg	61.55 ± 0.8	63.42 ± 1.0	67.22 ± 1.1	61.55 ± 0.8	50.60 ± 1.0	30.08 ± 1.3
LwF	Reg + R	61.55 ± 0.8	72.76 ± 2.9	**74.90 ± 2.9**	61.55 ± 0.8	31.77 ± 3.4	31.52 ± 3.2
CL VAE 5000	GR	61.55 ± 0.8	**73.12 ± 2.5**	66.86 ± 3.2	61.55 ± 0.8	67.97 ± 2.7	50.61 ± 3.1
BI-R	GR	61.55 ± 0.8	52.09 ± 2.4	52.70 ± 2.6	61.55 ± 0.8	22.65 ± 2.7	26.47 ± 2.4
Flashcards 5000	FR	61.55 ± 0.8	70.75 ± 1.0	72.13 ± 0.9	61.55 ± 0.8	**67.64 ± 1.0**	**63.71 ± 0.8**

Notes: When the task identifier is provided, flashcards' performance matches other baseline methods. However, in the absence of the task identifier, baselines fail, while flashcards help retain accuracy. Flashcards are robust to degradation across different domain tasks and outperform other methods. Reg: regularization, Reg + R: regularization and replay, ER = episodic replay, GR = generative replay, and FR = flashcard replay. Reported scores are averaged over three independent runs.

and an Adam optimizer with a learning rate of 0.0001 are used. Both the classifier and autoencoder models are trained for 5,000 iterations.

Results for Sequence3 (Cifar10–MNIST–Fashion MNIST) from Table 8.13 show that the baseline methods designed to make decisions only on seeing the task identifier during inference perform poorly in its absence. On the other hand, flashcards, by design, are task agnostic, and the network needs to rely only on the most recent checkpoint. It is competitive in the presence of the task identifier, and in its absence, it outperforms baseline methods by a big margin.

8.4.4 New instance classification

Single-task, new-instance learning (ST-NIL) classification introduced by Lomonaco and Maltoni (2017), focuses on learning new instances every session while retaining the same number of classes. Performance indicates how well the model adapts to virtual concept drift across sessions. Flashcards are constructed from the autoencoder per session and passed on to the classifier to get predicted softmax scores as soft class labels. It is observed that flashcards' performance is better than regularization and on par with episodic replay without explicitly storing exemplars in memory (Table 8.14).

Table 8.14: ST-NIL classification on Cifar10 using the settings described by Tao *et al.* (2020).

Session/method	Type	1	2	3	4	5
Naive*	—	67.80	69.31	71.37	73.12	73.23
Cumulative*	—	67.80	76.13	81.22	81.83	82.12
EWC*	Reg	67.80	69.45	72.68	74.02	74.31
SI*	Reg	67.80	70.48	72.82	74.63	74.58
IMM*	Reg	67.80	69.69	72.85	74.37	73.84
EEIL 1K*	ER	67.80	71.97	73.27	**74.91**	74.66
A-GEM 1K*	ER	67.80	**72.27**	**73.72**	74.81	**75.15**
Flashcards 1K	FR	67.90	71.40	73.34	74.84	74.88

Notes: *are reported from the same paper. Flashcards created from unsupervised AE perform equally well in comparison to other methods primarily built for classification. Reg: regularization, ER: episodic replay, FR: flashcard replay.

Table 8.15: ST-NIL classification on Cifar10 on the modified brightness and saturation setting with mean and standard deviation over five runs.

Method/Sess	1	2	3	4	5
Naive	67.9 ± 0.8	67.3 ± 0.8	67.6 ± 0.2	61.1 ± 0.9	63.0 ± 0.8
Cumulative	67.9 ± 0.8	69.7 ± 0.5	72.0 ± 0.3	72.8 ± 0.3	73.5 ± 0.2
Coreset 500	67.9 ± 0.8	67.5 ± 0.2	68.3 ± 0.1	61.9 ± 0.8	63.8 ± 0.1
Coreset 5000	67.9 ± 0.8	**68.4 ± 0.5**	68.3 ± 0.8	64.1 ± 0.4	65.1 ± 0.9
Flashcards	67.9 ± 0.8	68.1 ± 0.9	68.1 ± 1.2	**65.6 ± 1.3**	**65.8 ± 1.07**

ResNet18 is used as the classifier, optimized using SGD with learning rates of 0.001 over 20 epochs, and new sessions are introduced by adopting the same brightness and saturation adopted by Tao et al. (2020) (brightness_jitter = [0, −0.2, −0.1, 0.1, 0.2] and saturation_jitter = [0, 0.1, 0.2, −0.2, −0.1]), with the test set being constant across sessions. In order to introduce a more challenging setting across sessions, the following changes are made: brightness_jitter = [0, −0.1, 0.1, −0.2, 0.2] and saturation_jitter = [0, −0.1, 0.1, −0.2, 0.2], and results for the modified setting are provided in Table 8.15.

8.5 Concluding Thoughts

This chapter presented *flashcards* that could capture knowledge representations of a trained autoencoder by recursive passing of random image patterns and showed that they could be used as an alternative to original data for single tasks. Moreover, the reader would have observed its efficacy as a task-agnostic replay mechanism for various continual learning scenarios, including reconstruction, denoising, and task-incremental learning with heterogeneous datasets. Using flashcards for replay enables better performance compared to generative replay and regularization without requiring additional memory or training while also matching episodic replay performance without storing exemplars. By nature, flashcards allow for the abstraction of data, which can be exploited for potential data privacy applications. Generalization to other domains will foster further research in this area.

© 2024 World Scientific Publishing Company
https://doi.org/10.1142/9789811286711_0009

Chapter 9

Reliable AI-Based Decision Support System for Chest X-Ray Classification Using Continual Learning

Theivendiram Pranavan[*,‡] and ArulMurugan Ambikapathi[†,§]

[*]*National University of Singapore, Singapore*
[†]*Institute for Infocomm Research, Agency for Science, Technology and Research (A*STAR), Singapore*
[‡]*pranavan@u.nus.edu*
[§]*a.arulmurugan@gmail.com*

Abstract
This chapter discusses the challenges of machine learning models that need to learn from multiple sources due to regulations and confidentiality in practical domains such as security, finance and medical image analysis. This chapter presents a practical use case of continual learning using X-ray image classification in medical image analysis and demonstrates that continual learning is a helpful tool for data security through model sharing across multiple institutions/organizations. The chapter proposes an elastic weight consolidation (EWC)-based continual learning framework for X-ray image classification and an OOD detection framework using latent space features of the trained model. The proposed approach is experimentally validated using two X-ray image datasets, CheXpert and MIMIC-CXR, and provides a reliable and generalizable clinical decision support tool for X-ray image classification. The key elements of the study

consist of continual learning for handling forward data transfer and minimizing catastrophic forgetting, the data drift analysis method and OOD detection across datasets and an experimental validation of the proposed framework.

9.1 Introduction

Machine learning models that need to learn from multiple sources are a challenging but inevitable problem in many practical domains, such as security, finance, and medical image analysis. It comes with several intrinsic issues. The fundamental concern is that multiple sources may not be mutually sharing information due to regulations and confidentiality. For instance, in the security domain, a firm is unwilling to share their data/issues with other firms since the data can be potentially manipulated, which in turn may have a huge impact. In finance, confidential financial data are often considered critical when shared with another source. In medical image analysis, the patient data are not usually shared with third parties since the regulations for disclosure are often cumbersome owing to Personal Data Protection Act (PDPA) constraints (Wachter, 2018; Politou et al., 2018). On the other hand, learning from only one source may work poorly on another source or sometimes even be insufficient for the model, or knowledge learned from a new source may forget the knowledge learned from the old source (in the continual learning pretext, this is often referred to as *catastrophic forgetting*). In this chapter, we discuss a practical use case of continual learning (CL) using X-ray image classification in the domain of medical image analysis. We also demonstrate that CL is a very helpful tool for data security through model sharing across multiple institutions/organizations.

Artificial intelligence (AI) has been used in X-ray image processing, understanding, segmentation, and classification for several years (Catros and Mischler, 1988; Li et al., 1996; Davis et al., 1999; Deepa et al., 2011; Hassanien et al., 2020). The recent success of deep learning (DL) in image classification makes problem-solving easier for X-ray image classification. However, data sharing across several hospitals and care centers is a huge challenge due to local confidentiality and PDPA constraints, inhibiting the ability to develop broadly generalizable models across a diverse patient population. Furthermore, there is a drift in the data characteristics, either within the same care center or across care centers. Data collection time (the time between data gathering time and the actual testing), different sources, and data collection errors are common reasons for data drift. Hence, there is a need to detect drifts in the data distribution, as they occur, and to seek the clinician's assistance in decision-making, if required. Drift detection and model adaptation to the drift are required for the development of an objective clinical decision support tool that is generalizable across a larger patient population without defying the PDPA and confidentiality constraints. Federated learning is a way to train on different data sources locally without sharing the data globally across all the sources. Distributed learning is another way to learn a model from different sources. However, these learning paradigms do not facilitate how the model can continually learn across different sources without forgetting previously learned information.

In this chapter, we first present a CL framework based on elastic weight consolidation (EWC) (Kirkpatrick *et al.*, 2017) for X-ray image classification to design a clinical decision support tool that can generalize across the patient population, using data from two demographically different health care centers. Further, we also propose a support tool that is capable of isolating out-of-distribution (OOD) data, which helps to seek medical practitioners' assistance to aid in labeling/decision-making. We demonstrate the reliability and generalizability of the proposed approach using two X-ray image datasets (CheXpert (Irvin *et al.*, 2019) and MIMIC-CXR (Johnson *et al.*, 2019)).

Figures 9.1, 9.8 and 9.9 show the overall framework. In the overall framework, we feed the two datasets, and we continually learn a binary classifier to decide whether a particular disease is present or not. In certain inference cases where the machine learning model is not confident about the classification decision, the model seeks the clinician's assistance for expert labeling because the data are OOD. In the OOD detection framework, we learn the latent features from a deep neural network. For each latent feature, the model detects OOD. The final decision will be made using all N number of latent features. Sections 9.2 and 9.5 discuss more about the framework and OOD in detail. Although we use medical imaging datasets in our work, the framework is not limited to the medical domain as it can be generalized to other domains of interest. The following are the key contributions that are discussed in this chapter:

(1) CL for handling forward data transfer and minimizing catastrophic forgetting;

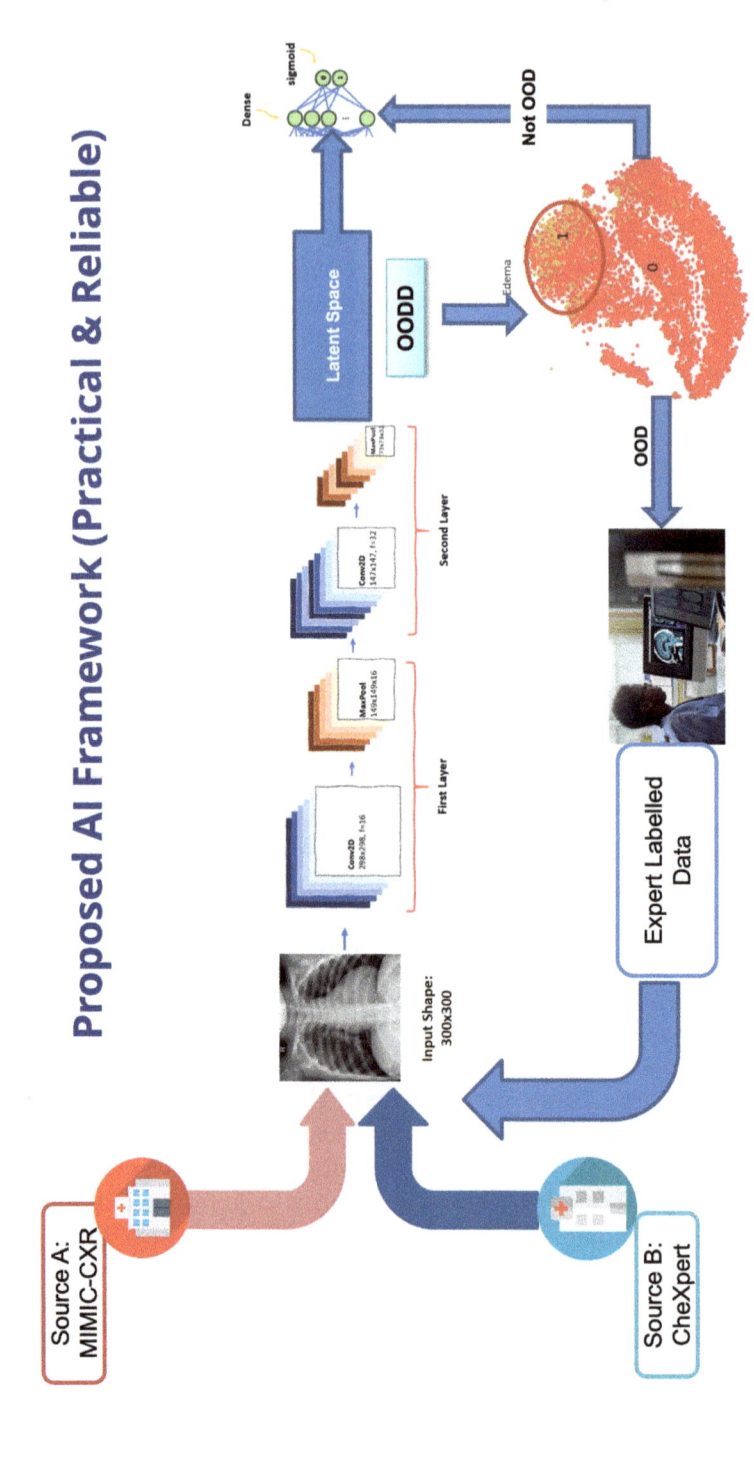

Fig. 9.1: Proposed AI framework: In this framework, we feed the given two datasets, and we continually learn a binary classification of whether a disease is there or not. In certain cases of inference where the machine learning model is not confident about the classification decision, the model seeks medical practitioner's assistance for expert labeling.

(2) data drift analysis method and OOD detection across datasets using latent space features of the trained model;
(3) experimental validation for the proposed *practical* and *reliable* frameworks for a decision support system.

9.2 Datasets and Problem Statement

In this work, we use two X-ray image datasets: CheXpert (Irvin *et al.*, 2019) and MIMIC-CXR (Johnson *et al.*, 2019). Table 9.1 shows the details of the datasets. Figures 9.2 and 9.4 depict sample images from CheXpert and MIMIC-CXR, respectively.

Table 9.1: CheXpert and MIMIC-CXR datasets summary.

	CheXpert	MIMIC-CXR
Number of images	223648	371858
Number of patients	64740	65079
Classes	14	14
Male	59.36%	52.17%
Female	40.64%	47.83%
0–20 (age)	0.87%	2.20%
20–40 (age)	13.18%	19.51%
40–60 (age)	31.00%	37.20%
60–80 (age)	38.94%	34.12%
80– (age)	16.01%	6.96%

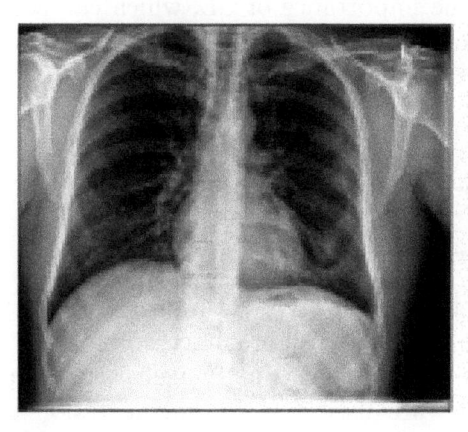

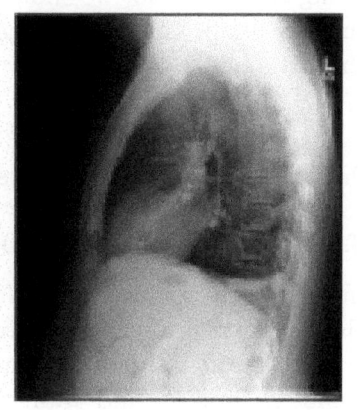

Fig. 9.2: CheXpert dataset: Sample images. Here, the two images show the frontal and lateral views of the same patient.

CheXpert images are categorized into 14 classes of pathological decisions. In addition, a specific image can take more than one class out of 14 classes. Four different labels are used for each class (Blank: Not mentioned, 0: negative, −1: uncertain, and 1: positive). MIMIC-CXR has the same classes, and a similar labeling scheme is used. In MIMIC-CXR, each imaging study contains two frontal and lateral views in most cases. However, both datasets originate from demographically different sources or healthcare centers. The former comes from Stanford Hospital, and the latter comes from Beth Israel Deaconess Medical Center. In this work, we choose four classes: edema, cardiomegaly, consolidation and pneumothorax (Fig. 9.3 shows some X-ray images for this learning problem). Each class is trained using a binary classification model. We do not use the whole dataset; we select a subset with disease and healthy labels. In addition, we chose only the four disease classes mentioned above for our training and testing. The data drift in CheXpert and MIMIC-CXR can be detected by their performance drops in cross-testing (see Section 9.3). Hence, the conventional machine learning pipeline of training and testing does not perform well on the new datasets with data drift. We show the ability of CL strategies ECW and L_2 to mitigate the drop in performance. In addition, we provide a preliminary way of detecting OOD. In this work, we first test performance on the individual dataset (i.e., the same dataset is divided into training and testing sets). Then, we train on one dataset and test on the other dataset to measure the cross-performance/intrinsic drift in the data. Using this scenario, we show the importance of CL, which can continually adapt the model to generalize to both datasets. Finally, we show the OOD detection for expert labeling.

9.3 Individual and Cross-Testing Performance

9.3.1 *Deep neural network training set up*

In this work, we use a ResNet-50-based architecture for all binary classification problems. Figure 9.5 shows the architecture.

ResNet-50 weights are pretrained from the ImageNet dataset. Random flipping and rotations are used for the augmentation of

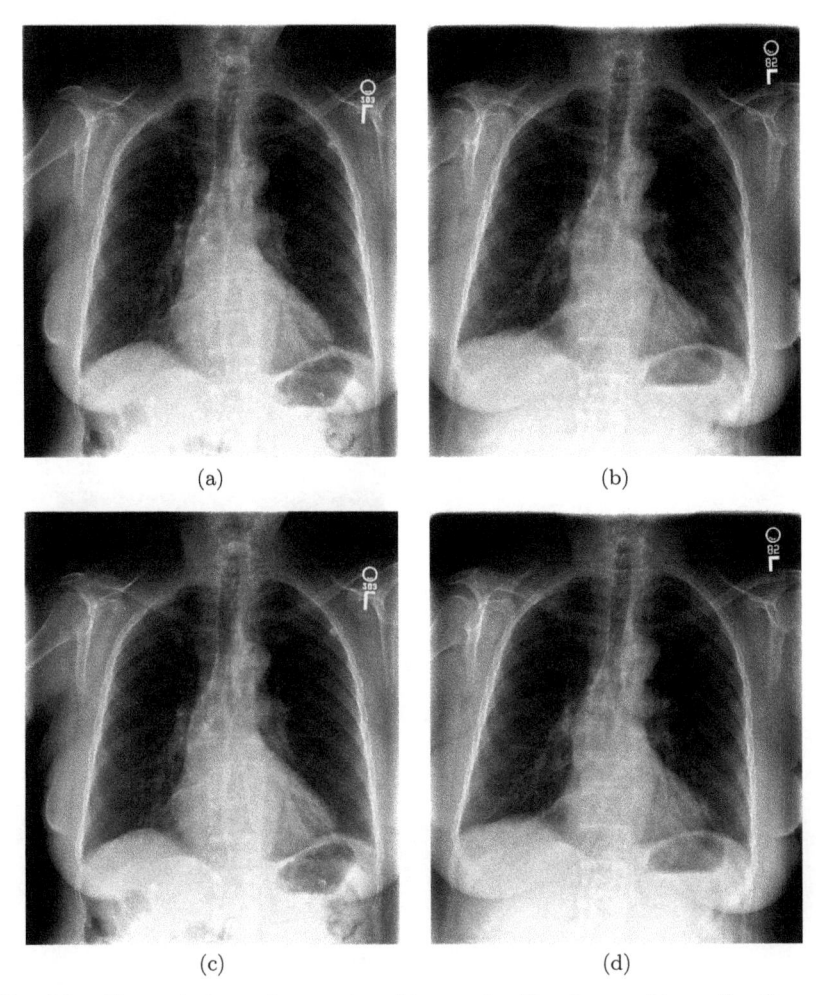

Fig. 9.3: Example input images in a binary classification problem of a disease: (a) and (b) images with the disease pneumothorax, and (c) and (d) are those without the disease.

the data. Table 9.2 shows various parameters used in the deep neural network. The models are trained to a maximum of 200 epochs with early stopping. The best models are selected based on the performance of the validation set. The performance is evaluated based on the AUC metric.

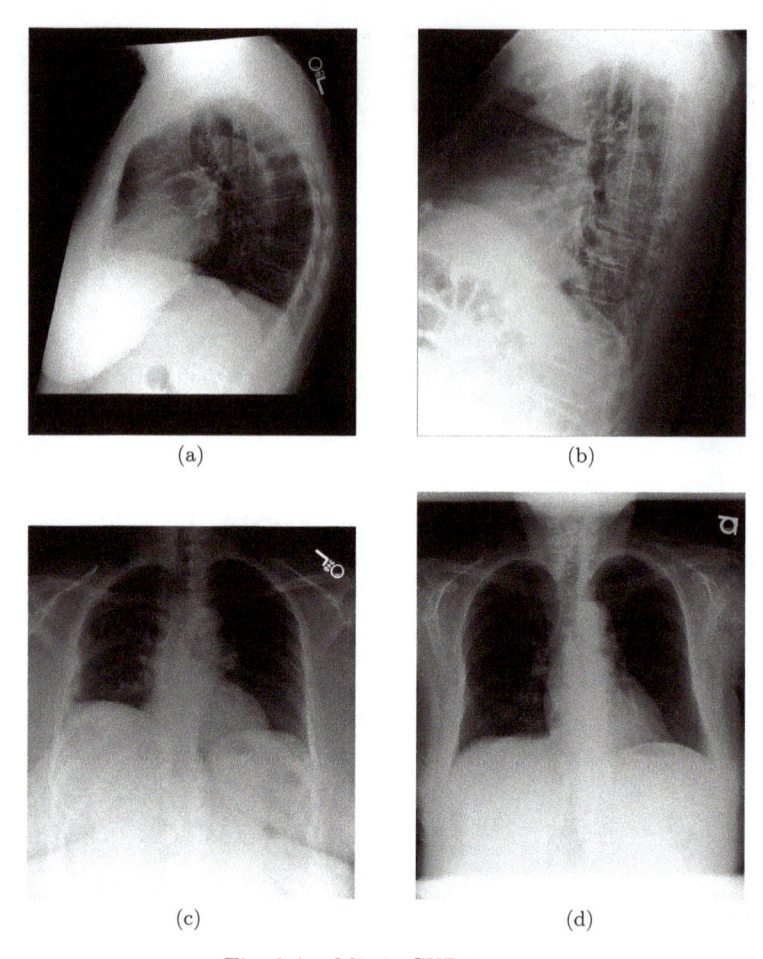

Fig. 9.4: Mimic-CXR images.

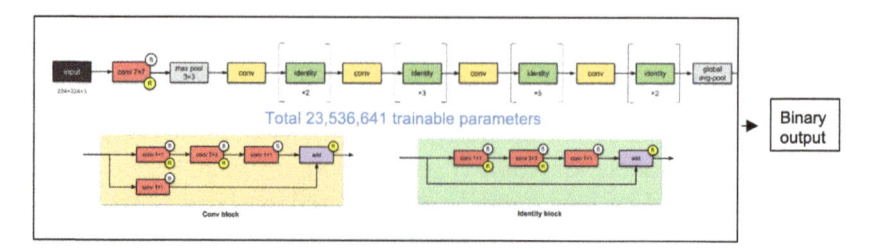

Fig. 9.5: ResNet-50-based architecture. The weights are pretrained from the Imagenet dataset. The final layer is modified to solve the binary classification problem. The same architecture is used for binary classification problems.

Table 9.2: Deep neural network settings.

Neural network parameter/variant	Experiment setting value/method
Batch size	128
Learning rate	$1e^{-4}$
Optimizer	Adam
Epochs	200 (Max)
Loss function	Binary cross-entropy
Callbacks	Early stopping

Table 9.3: CheXpert dataset performance.

Disease	AUC-ROC	Number of healthy samples in test	Number of samples with disease in test
Edema	0.94	189	45
Cardiomegaly	0.86	166	68
Consolidation	0.89	201	33
Pneumothorax	0.81	226	8

Table 9.4: MIMIC-CXR dataset performance.

Disease	AUC-ROC	Number of healthy samples in test	Number of samples with disease in test
Edema	0.87	510	959
Cardiomegaly	0.87	259	1,258
Consolidation	0.77	186	326
Pneumothorax	0.73	990	144

9.3.2 Individual and cross-testing results

Initially, the models are evaluated for individual performance on the same dataset. Tables 9.3 and 9.4 show this performance. We mean that the dataset is trained and tested on the same dataset as this individual performance.

Now, we train on one dataset and test on the other dataset. Tables 9.5 and 9.6 show this performance. In all diseases, there is either a drop in AUC-ROC or no improvement. For the same disease, the edema model trained on the CheXpert dataset has an AUC score of 0.94 for the same dataset. Conversely, the model

Table 9.5: MIMIC model tested on CheXpert.

Disease	AUC-ROC	AUC-ROC drop
Edema	0.89	0.05
Cardiomegaly	0.83	0.03
Consolidation	0.87	0.02
Pneumothorax	0.80	0.01

Table 9.6: CheXpert model tested on MIMIC.

Disease	AUC-ROC	AUC-ROC drop
Edema	0.85	0.02
Cardiomegaly	0.87	0
Consolidation	0.69	0.08
Pneumothorax	0.71	0.02

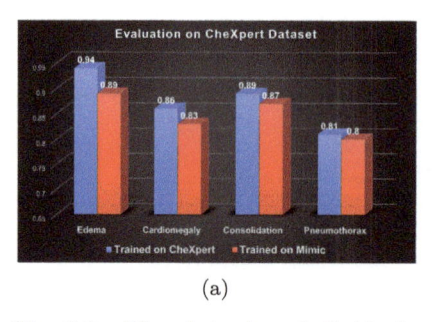

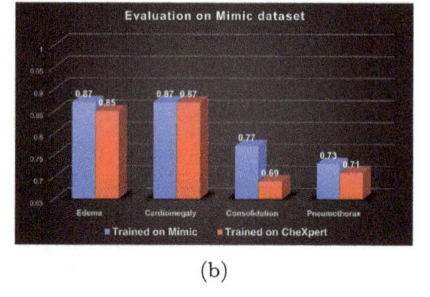

Fig. 9.6: The plots show individual and cross performance. The left plot (a) shows a model trained using the CheXpert dataset and evaluated on both datasets. Similarly, the plot on the right (b) is trained on MIMIC-CXR.

trained on MIMIC-CXR produced an AUC score of 0.89 on the CheXpert dataset. This is a drop of 0.05 in performance. Except for the CheXpert model for the disease cardiomegaly, all the other cross performances result in a drop in performance. This shows that there is a data drift in the datasets. Retraining from the whole dataset has two main problems. The first problem is that data sources have to share the dataset, which is impossible in most cases. The second is that retraining is often computationally expensive with growing data size. Hence, there is a need to learn continually from the multiple data sources. The overall results of individual and cross-testing performance are shown in Fig. 9.6.

9.4 CL with EWC and L_2

EWC is a way of learning a parameter set θ which can work for all tasks with an acceptable learning error. EWC assumes that the solution space with acceptable errors overlaps over several tasks. EWC proposes a way to find a solution in the overlapping region using selective regularization for all these tasks. For instance, we learn to task A first with a solution space, θ_A. When it comes to learning a new task B, the selective regularization penalizes important weights of θ_A more in task B. This strategy allows for learning a solution set, $\theta_{A,B}$, which is acceptable for both tasks. Figure 9.7 shows the overall idea of EWC with two tasks learned continually.

The optimization objective can be formulated as a Bayesian approach to estimate the parameter set θ given the dataset $\mathbb{D}$. We want to learn the posterior distribution, $p(\theta|\mathbb{D})$. In the case of two tasks A and B, $D = \{\mathbb{A}, \mathbb{B}\}$. With further manipulation using the Fisher information matrix, we have the following loss function (Eq. (9.1)) for learning the second task B after task A:

$$l(\theta) = l_B(\theta) - \frac{\lambda}{2}(\theta - \theta_A^\star)^\mathsf{T}\mathbb{I}_A(\theta - \theta_A^\star) + \epsilon' \quad (9.1)$$

Here, the loss is a summation of task B loss and a regularized loss of the previous task A, and ϵ' is a constant. The loss function can be extended to any number of tasks. The previous task is the learned parameter for all the previous tasks. The second learning method L_2 is conventional L_2 regularization.

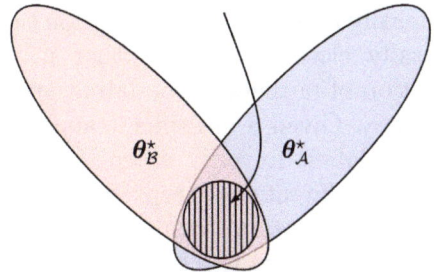

Fig. 9.7: EWC (Aich, 2021), which finds a parameter set in the overlapping region. It builds on the assumption that the solution spaces of various tasks overlap.

Table 9.7: CL strategy and results.

CheXpert → MIMIC		AUC-ROC				Gain/loss	
Disease	Data	Original	Transferred	L_2	EWC	L_2	EWC
Edema	CheXpert	0.94	0.94	0.94	0.94	0.00	0.00
Edema	MIMIC	0.87	**0.85**	0.89	0.91	+0.04	0.06
Cardiomegaly	CheXpert	0.86	0.86	0.89	0.87	+0.03	+0.01
Cardiomegaly	MIMIC	0.87	**0.87**	0.83	0.87	−0.04	0.00
Pneumothorax	CheXpert	0.81	0.81	0.82	0.81	+0.01	0.00
Pneumothorax	MIMIC	0.73	**0.71**	0.75	0.79	+0.04	+0.08

Note: The bolded values indicate the performance decline or maintenance when assessing a model trained on the ChestXpert Dataset and tested on MIMIC-CXR.

This section uses EWC and L_2 to learn and show the performance gain. Except for one case with the L_2 strategy, we see a performance gain or maintenance of the current performance when we learn continually. Table 9.7 shows the CL results. The model trained on CheXpert resulted in a drop in cross performance of 0.02 (0.87–0.85). However, both CL strategies L_2 and EWC gain 0.04 and 0.06 (Table 9.7, Edema, MIMIC row), respectively. EWC and L_2 gain or maintain the current performance in cross-testing except for one case when using the L_2 strategy, (Cardiomegaly, MIMIC).

9.5 OOD Data Analysis Based on Latent Space

In this section, we outline a preliminary way of performing OOD detection. It is primarily based on the assumption that in-distribution samples are generally closer to one another in the latent space, whereas the projection of outliers in the latent space dimension will be drifting or far away. Given a classifier trained on healthy versus non-healthy (disease) labels for only inlier samples, we first obtain the latent feature vector (latent space representations) for every input image. We then train two isolation forests (one for healthy and another for disease samples) to capture the two class-specific distributions. Each isolation forest predicts an anomaly score, which can be either positive or negative. The threshold of ≥ 0 on the anomaly score for a given input is considered an inlier. If a latent vector for an input when

Algorithm 9.1 OOD strategy: Training phase.

Require: $D_{\text{train}} = \{(\mathbf{X}_i, y_i)\}_{i=1}^{N}$: Set of N training images and corresponding ground-truth labels (either healthy or disease)

Require: $h(\mathbf{x};\ \theta^*)$: Pretrained DL model which takes an image as input and returns the latent space representations/activations

1: $S_{\text{healthy}}, S_{\text{disease}} \leftarrow \{\ \}, \{\ \}$, where S_{healthy} and S_{disease} denote the set of latent space representations for healthy and disease samples, respectively
2: **for** $(\mathbf{X}_i, y_i) \in D_{\text{train}}$ **do**
3:   **if** $y_i = healthy$ **then**
4:     $S_{\text{healthy}} \leftarrow S_{\text{healthy}} \cup \{h(\mathbf{X}_i;\ \theta^*)\}$
5:   **else**
6:     $S_{\text{disease}} \leftarrow S_{\text{disease}} \cup \{h(\mathbf{X}_i;\ \theta^*)\}$
7:   **end if**
8: **end for**
9: Train isolation forest models I_{healthy} and I_{disease} on S_{healthy} and S_{disease} training samples, respectively
10: **return** $I_{\text{healthy}}, I_{\text{disease}}$

passed to the two isolation forests yields negative anomaly scores for both (i.e., below the threshold of zero), we classify the input as OOD with respect to both healthy and disease samples, and the final decision for that input is determined to be OOD.

The formal training procedure for the approach discussed above is provided in Algorithm 9.1. Given a D_{train} training dataset with all samples being considered inliers (containing both healthy and non-healthy/disease patients) and a pretrained DL model that projects input training samples onto their latent space representations, obtaining sets S_{healthy} and S_{disease} for healthy and disease samples, respectively, we train two separate isolation forests, I_{healthy} and I_{disease}, independently using S_{healthy} and S_{disease}. A visual representation of the OOD training procedure is provided in Fig. 9.8. For inference, as formalized in Algorithm 9.2, we first obtain the latent space representation of the test image $\mathbf{X}_{\text{test}}$ as $\mathsf{ls}_{\text{test}}$ and pass it through the two isolation forests I_{healthy} and I_{disease}. Anomaly scores are computed for both cases as $\text{OOD}_{\text{healthy}}$ and $\text{OOD}_{\text{disease}}$. If both scores are negative (i.e., $<$), then it implies that the test

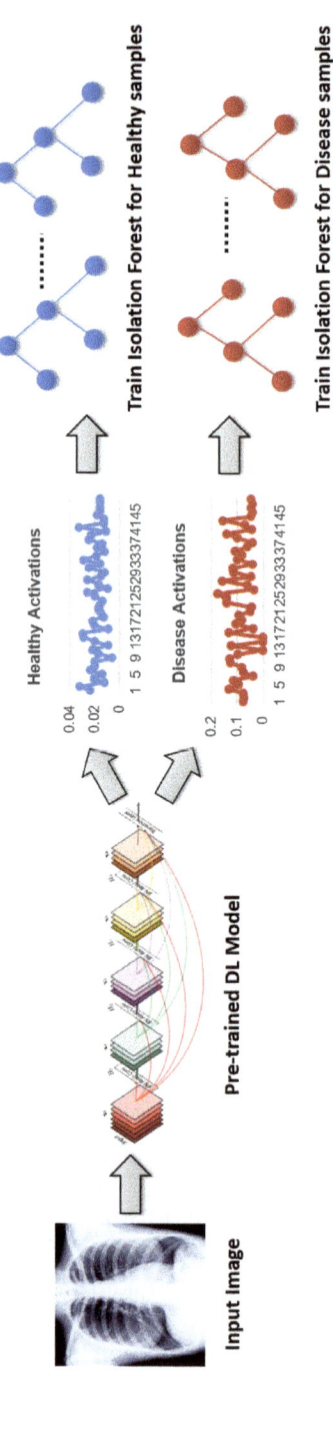

Fig. 9.8: OOD detection framework: Training. Given the set of training images, two isolation forest models are trained for OOD detection (as an anomaly case) over latent space representations obtained via a pretrained DL model.

Algorithm 9.2 OOD strategy: Inference phase.

Require: X_{test}: Test image
Require: $h(x; \theta^*)$: Pretrained DL model which takes an image as input and returns the latent space representations/activations
Require: $I_{healthy}(x)$, $I_{disease}(x)$: Isolation forest model for healthy and disease samples, which takes image as input and returns anomaly score

1: $ls_{test} \leftarrow h(X_{test}; \theta^*)$
2: $OOD_{healthy} \leftarrow I_{healthy}(ls_{test})$
3: $OOD_{disease} \leftarrow I_{disease}(ls_{test})$
4: **if** $OOD_{healthy} < 0$ and $OOD_{disease} < 0$ **then**
5:    **return** $True$
6: **else**
7:    **return** $False$
8: **end if**

---

sample latent representation is not part of either a healthy or non-healthy/disease distribution, thereby inferring it is likely to be an OOD sample. Figure 9.9 provides a visual representation of the OOD inference phase.

We provide some examples to showcase the outlier samples. To demonstrate this, we observe the response of the OOD framework, considering several outlier cases. For each case, we consolidate 5–10 images that reflect the particular scenario. Figure 9.10 shows lateral X-ray images, which the CL model misclassifies. However, our OOD detector finds all images as OOD and filters them to be further checked by a human (a doctor). The proposed OOD strategy is able to capture all 10 images which were misclassified by the CL/DL model. Figure 9.11 shows noisy images obtained by applying Gaussian blur and shot noise. The proposed OOD detector finds most of the anomalies (having low/high blur or shot noise), except for a few cases. Figure 9.12 shows the OOD detector performance from various geometric augmentations, such as horizontal/vertical flips and rotations. The OOD detector was able to find all the OOD samples arising from a horizontal flip and rotation by 20 degrees; however, for the vertical flip and rotation by 90 degrees case, the OOD detector could not find most of the anomalies. We postulate that these scenarios occur due to the use of augmented samples along with training set

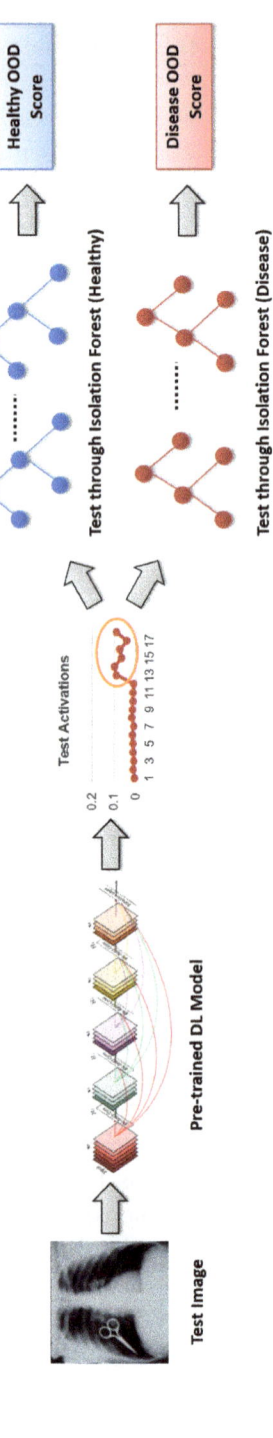

Fig. 9.9: OOD detection framework: Inference. The activations obtained from a pretrained DL model for a given test image is used to obtain anomaly scores from two trained (healthy and disease) isolation forest models, thereby determining whether the test image is OOD or not.

OOD Results – Lateral View

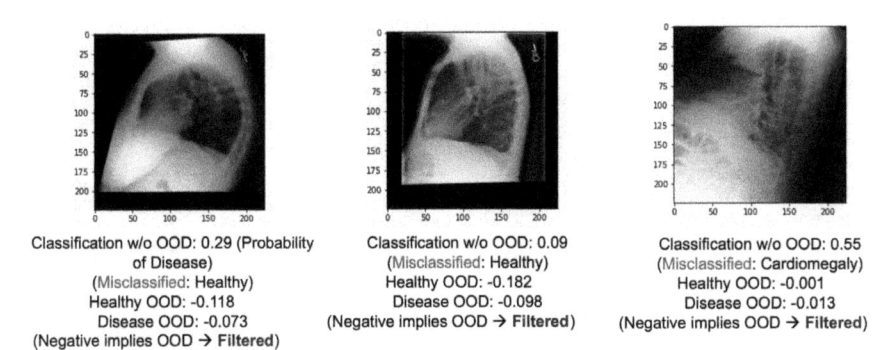

Classification w/o OOD: 0.29 (Probability of Disease)
(Misclassified: Healthy)
Healthy OOD: -0.118
Disease OOD: -0.073
(Negative implies OOD → Filtered)

Classification w/o OOD: 0.09
(Misclassified: Healthy)
Healthy OOD: -0.182
Disease OOD: -0.098
(Negative implies OOD → Filtered)

Classification w/o OOD: 0.55
(Misclassified: Cardiomegaly)
Healthy OOD: -0.001
Disease OOD: -0.013
(Negative implies OOD → Filtered)

10 / 10 test lateral images were filtered

Fig. 9.10: Lateral view images are fed to the OOD detector. The proposed OOD detector can detect all 10 samples which are misclassified as OOD. Here, if two OOD scores are negative, the sample does not belong to either of the distributions.

OOD Results – Noisy Images

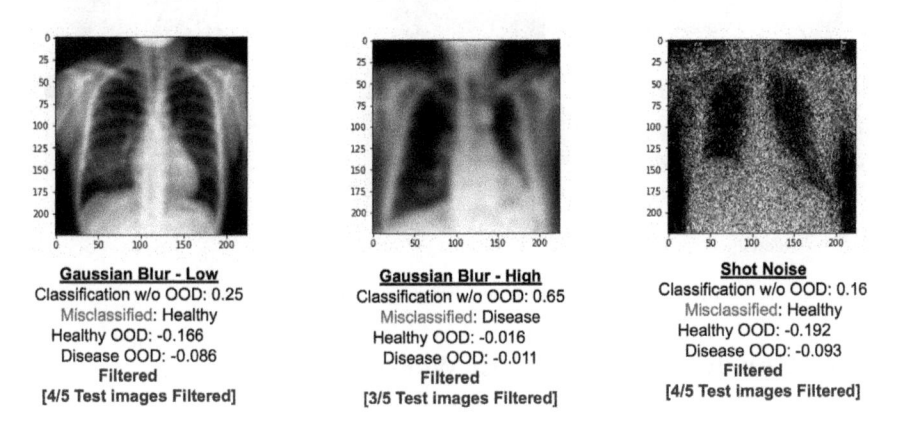

Gaussian Blur - Low
Classification w/o OOD: 0.25
Misclassified: Healthy
Healthy OOD: -0.166
Disease OOD: -0.086
Filtered
[4/5 Test images Filtered]

Gaussian Blur - High
Classification w/o OOD: 0.65
Misclassified: Disease
Healthy OOD: -0.016
Disease OOD: -0.011
Filtered
[3/5 Test images Filtered]

Shot Noise
Classification w/o OOD: 0.16
Misclassified: Healthy
Healthy OOD: -0.192
Disease OOD: -0.093
Filtered
[4/5 Test images Filtered]

Fig. 9.11: OOD detection with Gaussian blur and shot noise.

samples while training DL/CL models. Moreover, our model is able to detect images that are completely different from X-ray images, as shown in Fig. 9.13. Hence, our model is able to learn the distributions of healthy and non-healthy X-ray images and detect OOD samples.

OOD Results – Noisy Images

Horizontal Flip
Classification w/o OOD: 0.22
Misclassified: Healthy
Healthy OOD: -0.108
Disease OOD: -0.082
Filtered
[5/5 Test images Filtered]

Vertical Flip
Classification w/o OOD: 0.10
Misclassified: Healthy
Healthy OOD: -0.173
Disease OOD: -0.085
Filtered
[2/5 Test images Filtered]

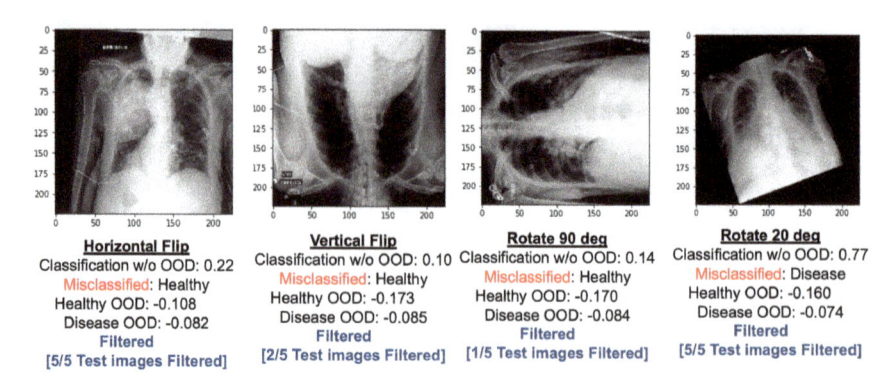

Rotate 90 deg
Classification w/o OOD: 0.14
Misclassified: Healthy
Healthy OOD: -0.170
Disease OOD: -0.084
Filtered
[1/5 Test images Filtered]

Rotate 20 deg
Classification w/o OOD: 0.77
Misclassified: Disease
Healthy OOD: -0.160
Disease OOD: -0.074
Filtered
[5/5 Test images Filtered]

Fig. 9.12: OOD with augmentation.

OOD Results – Special Mentions

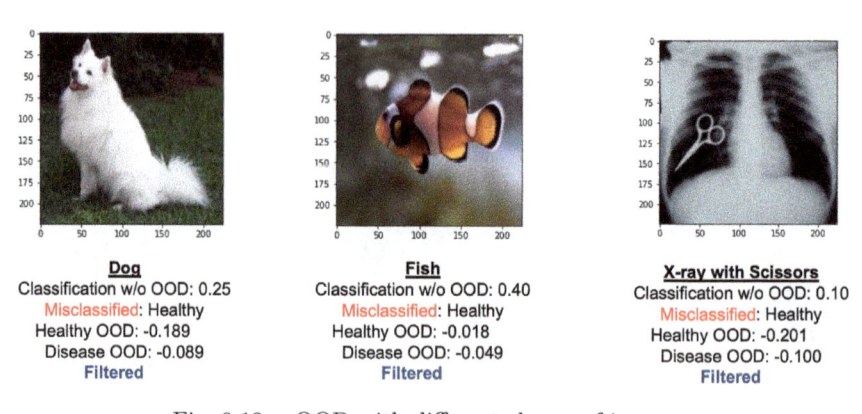

Dog
Classification w/o OOD: 0.25
Misclassified: Healthy
Healthy OOD: -0.189
Disease OOD: -0.089
Filtered

Fish
Classification w/o OOD: 0.40
Misclassified: Healthy
Healthy OOD: -0.018
Disease OOD: -0.049
Filtered

X-ray with Scissors
Classification w/o OOD: 0.10
Misclassified: Healthy
Healthy OOD: -0.201
Disease OOD: -0.100
Filtered

Fig. 9.13: OOD with different classes of images.

9.6 Conclusion and Future Works

Learning from multiple sources from different data sources is challenging due to the various regulations of the data sources. Essentially, the sources will not be sharing the data with another third party. Hence, there is a need to develop machine learning models that can continually learn without forgetting what they have learned previously. This chapter presented a CL approach using EWC and L_2 to handle data drift across two X-ray image datasets:

CheXpert and MIMIC-CXR. The existing data drift among the given datasets results in a drop in cross/transfer performance. The proposed approach maintains or gains performance except for one experiment. Moreover, we presented a way to detect OOD data that need to be labeled by experts. Our proposed OOD detector can detect OOD in the same domain and an image that is entirely out of the domain.

We focus on following future directions as the next step:

- considering all 14 classes or more across datasets;
- strengthen fine-tuning strategies for both OOD detections and CL;
- rigorous evaluations (along with other CXR datasets/classes as well).

Acknowledgments

We thank all the group members (Gopalakrishnan Saisubramaniam, Pranshu Ranjan Singh, Guo Yang, Cheryl Wong Sze Yin, Wang Jie, Ramanpreet Singh Pahwa, and Savitha Ramasamy) who contributed to the work.

Bibliography

Abati, D., Tomczak, J., Blankevoort, T., Calderara, S., Cucchiara, R., and Bejnordi, B. E. (2020). Conditional channel gated networks for task-aware continual learning, in *IEEE/CVF Conference on Computer Vision and Pattern Recognition*, pp. 3931–3940.

Aich, A. (2021). Elastic weight consolidation (EWC): Nuts and bolts. *arXiv preprint* arXiv:2105.04093.

Akata, Z., Perronnin, F., Harchaoui, Z., and Schmid, C. (2013). Label-embedding for attribute-based classification, in *Proceedings of the IEEE Conference on Computer Vision and Pattern Recognition*, pp. 819–826.

Akata, Z., Reed, S., Walter, D., Lee, H., and Schiele, B. (2015). Evaluation of output embeddings for fine-grained image classification, in *Proceedings of the IEEE Conference on Computer Vision and Pattern Recognition*, pp. 2927–2936.

Al-Shedivat, M., Bansal, T., Burda, Y., Sutskever, I., Mordatch, I., and Abbeel, P. (2018). Continuous adaptation via meta-learning in non-stationary and competitive environments, in *International Conference on Learning Representations (ICLR)*.

Aljundi, R., Chakravarty, P., and Tuytelaars, T. (2017). Expert gate: Lifelong learning with a network of experts, in *Proceedings of the IEEE Conference on Computer Vision and Pattern Recognition*, pp. 3366–3375.

Aljundi, R., Babiloni, F., Elhoseiny, M., Rohrbach, M., and Tuytelaars, T. (2018). Memory aware synapses: Learning what (not) to forget, in *Proceedings of the European Conference on Computer Vision (ECCV)*, pp. 139–154.

Aljundi, R., Caccia, L., Belilovsky, E., Caccia, M., Lin, M., and Charlin, L. (2019a). Online continual learning with maximally interfered retrieval, in *Proceedings of the Advances in Neural Information Processing Systems*, Vol. 32.

Aljundi, R., Kelchtermans, K., and Tuytelaars, T. (2019b). Task-free continual learning, in *IEEE/CVF Conference on Computer Vision and Pattern Recognition*, pp. 11254–11263.

Amari, S.-I. (1998). Natural gradient works efficiently in learning, *Neural Computation* **10**(2), 251–276.

Annadani, Y. and Biswas, S. (2018). Preserving semantic relations for zero-shot learning, in *Proceedings of the IEEE Conference on Computer Vision and Pattern Recognition*, pp. 7603–7612.

Arjovsky, M., Chintala, S., and Bottou, L. (2017). Wasserstein gan, **30**. *arXiv preprint* arXiv:1701.07875.

Baesens, B., Gestel, T. V., Viaene, S., Stepanova, M., Suykens, J., and Vanthienen, J. (2003). Benchmarking state-of-the-art classification algorithms for credit scoring, *The Journal of the Operational Research Society* **54**(6), 627–635.

Bang, J., Kim, H., Yoo, Y., Ha, J.-W., and Choi, J. (2021). Rainbow memory: Continual learning with a memory of diverse samples, in *IEEE/CVF Conference on Computer Vision and Pattern Recognition*, pp. 8218–8227.

Belouadah, E. and Popescu, A. (2019). Il2m: Class incremental learning with dual memory, in *Proceedings of the IEEE/CVF International Conference on Computer Vision*, pp. 583–592.

Bengio, Y. and Delalleau, O. (2009). Justifying and generalizing contrastive divergence, *Neural Computation* **21**(6), 1601–1621.

Bing, L. (2020). Learning on the job: Online lifelong and continual learning, in *Proceedings of the AAAI*.

Blundell, C., Cornebise, J., Kavukcuoglu, K., and Wierstra, D. (2015). Weight uncertainty in neural network, in *International Conference on Machine Learning*.

Borsos, Z., Mutny, M., and Krause, A. (2020). Coresets via bilevel optimization for continual learning and streaming, in *Advances in Neural Information Processing Systems*.

Bouchard, G., Trouillon, T., Perez, J., and Galdon, A. (2016). Online learning to sample. *arXiv preprint* arXiv:1506.09016v2.

Broderick, T., Boyd, N., Wibisono, A., Wilson, A. C., and Jordan, M. I. (2013). Streaming variational Bayes, in *Advances in Neural Information Processing Systems*.

Buch, E. R., Claudino, L., Quentin, R., Bönstrup, M., and Cohen, L. G. (2021). Consolidation of human skill linked to waking hippocampo-neocortical replay, *Cell Reports* **35**(10), 109193. https://doi.org/10.1016/j.celrep.2021.109193.

Cacheux, Y. L., Borgne, H. L., and Crucianu, M. (2019). Modeling inter and intra-class relations in the triplet loss for zero-shot learning, in *Proceedings of the IEEE/CVF International Conference on Computer Vision*, pp. 10333–10342.

Castro, F. M., Marín-Jiménez, M. J., Guil, N., Schmid, C., and Alahari, K. (2018). End-to-end incremental learning, in *Proceedings of the European Conference on Computer Vision (ECCV)*, pp. 233–248.

Catros, J.-Y. and Mischler, D. (1988). An artificial intelligence approach for medical picture analysis. *Pattern Recognition Letters* **8**(2), 123–130.

Changpinyo, S., Chao, W.-L., and Sha, F. (2017). Predicting visual exemplars of unseen classes for zero-shot learning, in *Proceedings of the IEEE International Conference on Computer Vision*, pp. 3476–3485.

Chaudhry, A., Dokania, P. K., Ajanthan, T., and Torr, P. H. (2018a). Riemannian walk for incremental learning: Understanding forgetting and intransigence, in *Proceedings of the European Conference on Computer Vision (ECCV)*, pp. 532–547.

Chaudhry, A., Ranzato, M., Rohrbach, M., and Elhoseiny, M. (2018b). Efficient lifelong learning with A-GEM, in *International Conference on Learning Representations*.

Chaudhry, A., Rohrbach, M., Elhoseiny, M., Ajanthan, T., Dokania, P. K., Torr, P. H., and Ranzato, M. (2019). On tiny episodic memories in continual learning. *arXiv preprint* arXiv:1902.10486.

Chen, G., Xu, R., and Srihari, S. (2015). Sequential labeling with online deep learning. *arXiv preprint* arXiv:1412:3397v3.

Chen, Y., Diethe, T., and Lawrence, N. (2018). Facilitating Bayesian continual learning by natural gradients and Stein gradients, in *Continual Learning Workshop @ NeurIPS*.

Chen, P.-H., Wei, W., Hsieh, C.-J., and Dai, B. (2021). Overcoming catastrophic forgetting by Bayesian generative regularization, in *International Conference on Machine Learning*.

Chik, W. B. (2013). The Singapore Personal Data Protection Act and an assessment of future trends in data privacy. *Computer Law and Security Review*, **29**(5), 554–575.

Chou, Y.-Y., Lin, H.-T., and Liu, T.-L. (2021). Adaptive and generative zero-shot learning, in *International Conference on Learning Representations*.

Cossu, A., Graffieti, G., Pellegrini, L., Maltoni, D., Bacciu, D., Carta, A., and Lomonaco, V. (2022a). Is class-incremental enough for continual learning? *Frontiers in Artificial Intelligence* **5**.

Cossu, A., Tuytelaars, T., Carta, A., Passaro, L., Lomonaco, V., and Bacciu, D. (2022b). Continual pre-training mitigates forgetting in language and vision. *arXiv preprint* arXiv:2205.09357.

Cote, M. A. and Larochelle, H. (2016). An infinite restricted Boltzmann machine. arXiv:1502.02476v4.

Danker, J. F. and Anderson, J. R. (2010). The ghosts of brain states past: Remembering reactivates the brain regions engaged during encoding, *Psychological Bulletin* **136**(1), 87.

Davis, D., Linying, S., and Sharp, B. (1999). Neural networks for X-ray image segmentation, in *Proceedings of the First International Conference on Enterprise Information Systems*. Escola superior de tecnologia, Setubal, Portugal, pp. 264–271.

Deepa, S. N. and Devi, B. A. (2011). A survey on artificial intelligence approaches for medical image classification. *Indian Journal of Science and Technology* **4**(11), 1583–1595.

Delange, M., Aljundi, R., Masana, M., Parisot, S., Jia, X., Leonardis, A., Slabaugh, G., and Tuytelaars, T. (2021). A continual learning survey: Defying forgetting in classification tasks, *IEEE Transactions on Pattern Analysis and Machine Intelligence*, 1. doi: 10.1109/TPAMI.2021.3057446.

Deng, J., Dong, W., Socher, R., Li, L.-J., Li, K., and Fei-Fei, L. (2009). ImageNet: A large-scale hierarchical image database, in *IEEE Conference on Computer Vision and Pattern Recognition*, pp. 248–255.

Devlin, J., Chang, M. W., Lee, K., and Toutanova, K. (2019). BERT: Pre-training of deep bidirectional transformers for language understanding, in *NAACL-HLT*.

Dhar, P., Singh, R. V., Peng, K.-C., Wu, Z., and Chellappa, R. (2019). Learning without memorizing, in *Proceedings of the IEEE/CVF Conference on Computer Vision and Pattern Recognition (CVPR)*, pp. 5138–5146.

Díaz-Rodríguez, N., Lomonaco, V., Filliat, D., and Maltoni, D. (2018). Don't forget, there is more than forgetting: New metrics for continual learning, in *Workshop on Continual Learning, NeurIPS 2018 (Neural Information Processing Systems)*, Montreal, Canada. https://hal.arc hives-ouvertes.fr/hal-01951488.

Dietterich, T. G. (2002). Chapter: Machine learning for sequential data: A review. *Structure, Synctactic and Statistical Pattern Recognition.* Berlin, Heidelberg: Springer-Verlag. pp. 15–30. https://doi.org/10.1007/3-540-70659-3_2

Draelos, T. J., Miner, N. E., Lamb, C. C., Cox, J. A., Vineyard, C. M., Carlson, K. D., Severa, W. M., James, C. D., and Aimone, J. B. (2016). Neurogenesis deep learning. *arXiv preprint* arXiv:1612.03770.

Ebrahimi, S., Elhoseiny, M., Darrell, T., and Rohrbach, M. (2019). Uncertainty-guided continual learning with Bayesian neural networks, in *International Conference on Learning Representations (ICLR)*.

Ebrahimi, S., Elhoseiny, M., Darrell, T., and Rohrbach, M. (2020a). Uncertainty-guided lifelong learning in Bayesian networks, in *International Conference on Learning Representations*.

Ebrahimi, S., Meier, F., Calandra, R., Darrell, T., and Rohrbach, M. (2020b). Adversarial continual learning, in *European Conference on Computer Vision (ECCV)*, Glasgow, UK, pp. 386–402.

Egorov, E., Kuzina, A., and Burnaev, E. (2021). BooVAE: Boosting approach for continual learning of VAE, in *Advances in Neural Information Processing Systems*, Vol. 34.

Elhoseiny, M., Zhu, Y., Zhang, H., and Elgammal, A. (2017). Link the head to the "beak": Zero shot learning from noisy text description at part precision, in *Proceedings of the IEEE Conference on Computer Vision and Pattern Recognition*, pp. 5640–5649.

Eschenhagen, R. (2019). Natural gradient variational inference for continual learning in deep neural networks, Technical report, University of Osnabruck.

Ester, M., Kriegel, H.-P., Sander, J., and Xu, X. (1996). A density-based algorithm for discovering clusters in large spatial databases with noise, in *Proceedings of 2nd International Conference on Knowledge Discovery and Data Mining (KDD-96)*, Vol. 96, pp. 226–231.

Farhadi, A., Endres, I., Hoiem, D., and Forsyth, D. (2009). Describing objects by their attributes, in *2009 IEEE Conference on Computer Vision and Pattern Recognition*. IEEE, pp. 1778–1785.

Farquhar, S. and Gal, Y. (2018). A unifying Bayesian view of continual learning, in *Bayesian Deep Learning Workshop @ NeurIPS*.

Fayek, H. M. (2019). Continual Deep Learning via Progressive Learning, Ph.D. thesis, RMIT University, Australia.

Fayek, H. M., Cavedon, L., and Wu, H. R. (2020). Progressive learning: A deep learning framework for continual learning, *Neural Networks* **128**, 345–357.

Felix, R., Kumar, V. B., Reid, I., and Carneiro, G. (2018). Multi-modal cycle-consistent generalized zero-shot learning, in *Proceedings of the European Conference on Computer Vision (ECCV)*, pp. 21–37.

French, R. M. (1999). Catastrophic forgetting in connectionist networks, *Trends in Cognitive Sciences* **3**(4), 128–135.

Frome, A., Corrado, G. S., Shlens, J., Bengio, S., Dean, J., Ranzato, M., and Mikolov, T. (2013). Devise: A deep visual-semantic embedding model, *Advances in Neural Information Processing Systems*, **26**, 2121–2129.

Geirhos, R., Rubisch, P., Michaelis, C., Bethge, M., Wichmann, F. A., and Brendel, W. (2019). Imagenet-trained CNNs are biased towards texture; increasing shape bias improves accuracy and robustness, in *ICLR*.

Ghahramani, Z. and Attias, H. (2000). Online variational Bayesian learning, in *NIPS Workshop on Online Learning*.

Ghosh, P. and Davis, L. S. (2018). Understanding center loss based network for image retrieval with few training data, in *Proceedings of the European Conference on Computer Vision (ECCV)*.

Gideon, J., Khorram, S., Aldeneh, Z., Dimitriadis, D., and Provost, E. M. (2017). Progressive neural networks for transfer learning in emotion recognition. *arXiv preprint* arXiv:1706.03256.

Golkar, S., Kagan, M., and Cho, K. (2019). Continual learning via neural pruning. *NeurIPS Workshop on Real Neurons & Hidden Units*, December 2019, Vancouver, British Columbia, Canada. *arXiv preprint* arXiv:1903.04476.

Goodfellow, I. J., Mirza, M., Xiao, D., Courville, A., and Bengio, Y. (2013). An empirical investigation of catastrophic forgetting in gradient-based neural networks, in *International Conference on Learning Representations*. arXiv:1312.6211 [stat.ML].

Goodfellow, I. J., Pouget-Abadie, J., Mirza, M., Xu, B., Warde-Farley, D., Ozair, S., Courville, A., and Bengio, Y. (2014). Generative adversarial nets, in *Advances in Neural Information Processing Systems*, pp. 2672–2680.

Gopalakrishnan, S., Singh, P. R., Fayek, H., Ramasamy, S., and Ambikapathi, A. (2020). Knowledge capture and replay for continual learning. *arXiv preprint* arXiv:2012.06789.

Gou, J., Yu, B., Maybank, S. J., and Tao, D. (2020). Knowledge distillation: A survey. arXiv:2006.05525 [cs.LG].

Graffieti, G., Maltoni, D., Pellegrini, L., and Lomonaco, V. (2022). Generative negative replay for continual learning. *arXiv preprint* arXiv:2204.05842.

Han, S., Meng, Z., Khan, A. S., and Tong, Y. (2016). Incremental boosting convolutional neural network for facial action unit recognition, *Neural Information Processing Systems*.

Hand, D. J. and Henley, W. E. (1997). Statistical classification models in consumer credit scoring: A review, *Journal of the Royal Statistical Society: Series A (General)* **160**, 523–541.

Hassanien, A. E., Mahdy, L. N., Ezzat, K. A., Elmousalami, H. H., and Ella, H. A. (2020). Automatic X-ray COVID-19 lung image classification system based on multi-level thresholding and support vector machine. *medRxiv*. Cold Spring Harbor Laboratory Press.

Hayes, T. L., Cahill, N. D., and Kanan, C. (2018a). Memory efficient experience replay for streaming learning, in *2019 International Conference on Robotics and Automation (ICRA)*, pp. 9769–9776.

Hayes, T. L., Kemker, R., Cahill, N. D., and Kanan, C. (2018b). New metrics and experimental paradigms for continual learning, in *2018 IEEE/CVF Conference on Computer Vision and Pattern Recognition Workshops (CVPRW)*, pp. 2112–21123. doi: 10.1109/CVPRW.2018.00273.

He, K., Zhang, X., Ren, S., and Sun, J. (2016). Deep residual learning for image recognition, in *Proceedings of the IEEE Conference on Computer Vision and Pattern Recognition*, pp. 770–778.

He, X., Sygnowski, J., Galashov, A., Rusu, A. A., Teh, Y. W., and Pascanu, R. (2020). Task agnostic continual learning via meta learning, in *Lifelong Machine Learning Workshop @ ICML*. arXiv. /abs/1906.05201.

Heusel, M., Ramsauer, H., Unterthiner, T., Nessler, B., and Hochreiter, S. (2017). Gans trained by a two time-scale update rule converge to a local Nash equilibrium, in *Advances in Neural Information Processing Systems*, pp. 6626–6637.

Hinton, G. E. (2002). Training products of experts by minimizing contrastive divergence, *Neural Computation* **14**(8), 1771–1800.

Hinton, G. E., Vinyals, O., and Dean, J. (2015). Distilling the knowledge in a neural network. arXiv:1503.02531, https://arxiv.org/abs/1503.02531.

Hou, S., Pan, X., Loy, C. C., Wang, Z., and Lin, D. (2018). Lifelong learning via progressive distillation and retrospection, in *European Conference on Computer Vision (EECV)*, pp. 452–467.

Hou, S., Pan, X., Loy, C. C., Wang, Z., and Lin, D. (2019). Learning a unified classifier incrementally via rebalancing, in *Proceedings of the IEEE/CVF Conference on Computer Vision and Pattern Recognition*, pp. 831–839.

Huang, G., Liu, Z., Van Der Maaten, L., and Weinberger, K. Q. (2017). Densely connected convolutional networks, in *Proceedings of the IEEE Conference on Computer Vision and Pattern Recognition*, pp. 4700–4708.

Huang, H., Wang, C., Yu, P. S., and Wang, C.-D. (2019). Generative dual adversarial network for generalized zero-shot learning, in *Proceedings of the IEEE Conference on Computer Vision and Pattern Recognition*, pp. 801–810.

Irvin, J., Rajpurkar, P., Ko, M., Yu, Y., Ciurea-Ilcus, S., Chute, C., Marklund, H., Haghgoo, B., Ball, R., Shpanskaya, K., Seekins, J., Mong, D. A., Halabi, S. S., Sandberg, J. K., Jones, R., Larson, D. B., Langlotz, C. P., Patel, B. N., Lungren, M. P., and Ng, A. Y. (2019).

Chexpert: A large chest radiograph dataset with uncertainty labels and expert comparison, in *33rd AAAI Conference on Artificial Intelligence, AAAI 2019, 31st Innovative Applications of Artificial Intelligence Conference, IAAI 2019 and the 9th AAAI Symposium on Educational Advances in Artificial Intelligence, EAAI* 2019, pp. 590–597.

Isele, D. and Cosgun, A. (2018). Selective experience replay for lifelong learning, in *Proceedings of the Thirty-Second AAAI Conference on Artificial Intelligence and Thirtieth Innovative Applications of Artificial Intelligence Conference and Eighth AAAI Symposium on Educational Advances in Artificial Intelligence (AAAI'18/IAAI'18/EAAI'18)*. AAAI Press, 404, 3302–3309.

Janowsky, S. A. (1989). Pruning versus clipping in neural networks, *Physical Review A* **39**(12).

Jerfel, G., Grant, E., Griffiths, T., and Heller, K. A. (2019). Reconciling meta-learning and continual learning with online mixtures of tasks, in *Advances in Neural Information Processing Systems*, Vol. 32.

Jin, X., Sadhu, A., Du, J., and Ren, X. (2021). Gradient-based editing of memory examples for online task-free continual learning, in *Advances in Neural Information Processing Systems*, Vol. 34.

Johnson, A. E. W., Pollard, T. J., Berkowitz, S. J., Greenbaum, N. R., Lungren, M. P., Deng, C.-y., Mark, R. G., and Horng, S. (2019). MIMIC-CXR, a de-identified publicly available database of chest radiographs with free-text reports, *Scientific Data* **6**(1), 1–8.

Kemker, R. and Kanan, C. (2018). Fearnet: Brain-inspired model for incremental learning, in *International Conference on Learning Representations*. https://openreview.net/forum?id=SJ1Xmf-Rb.

Kemker, R., McClure, M., Abitino, A., Hayes, T. L., and Kanan, C. (2017). Measuring catastrophic forgetting in neural networks, in *AAAI*.

Keshari, R., Singh, R., and Vatsa, M. (2020). Generalized zero-shot learning via over-complete distribution, in *Proceedings of the IEEE/CVF Conference on Computer Vision and Pattern Recognition*, pp. 13300–13308.

Kessler, S., Nguyen, V., Zohren, S., and Roberts, S. J. (2021). Hierarchical Indian buffet neural networks for Bayesian continual learning, in *Uncertainty in Artificial Intelligence*, pp. 749–759.

Khan, M., Nielsen, D., Tangkaratt, V., Lin, W., Gal, Y., and Srivastava, A. (2018). Fast and scalable Bayesian deep learning by weight-perturbation in Adam, in *International Conference on Machine Learning*.

Kingma, D. P. and Ba, J. (2015). Adam: A method for stochastic optimization, in *International Conference on Learning Representations*.

Kingma, D. P. and Welling, M. (2013). Auto-encoding variational Bayes. *arXiv preprint* arXiv:1312.6114.

Kingma, D. P. and Welling, M. (2014). Stochastic gradient VB and the variational auto-encoder, in *International Conference on Learning Representations*.

Kingma, D. P., Salimans, T., and Welling, M. (2015). Variational dropout and the local reparameterization trick, in *Advances in Neural Information Processing Systems*.

Kirkpatrick, J., Pascanu, R., Rabinowitz, N., Veness, J., Desjardins, G., Rusu, A. A., Milan, K., Quan, J., Ramalho, T., Grabska-Barwinska, A., Hassabis, D., Clopath, C., Kumaran, D., and Hadsell, R. (2017). Overcoming catastrophic forgetting in neural networks, *Proceedings of the National Academy of Sciences* **114**(13), 3521–3526.

Kiyasseh, D., Zhu, T., and Clifton, D. A. (2020). Clops: Continual learning of physiological signals. *arXiv preprint* arXiv:2004.09578.

Kodirov, E., Xiang, T., and Gong, S. (2017). Semantic autoencoder for zero-shot learning, in *Proceedings of the IEEE Conference on Computer Vision and Pattern Recognition*, pp. 3174–3183.

Krizhevsky, A. (2009). Learning multiple layers of features from tiny images, Technical report, University of Toronto, Toronto, Ontario.

Krizhevsky, A., Sutskever, I., and Hinton, G. E. (2012). Imagenet classification with deep convolutional neural networks, in *Advances in Neural Information Processing Systems*, Vol. 25, pp. 1097–1105.

Kullback, S. and Leibler, R. A. (1951). On information and sufficiency, *The Annals of Mathematical Statistics* **22**(1), 79–86.

Kumar, A., Chatterjee, S., and Rai, P. (2021). Bayesian structural adaptation for continual learning, in *International Conference on Machine Learning*.

Kumar, N., Savitha, R., and Mamun, A. (2018). Ocean wave height prediction using ensemble of extreme learning machine, *Neurocomputing* **277**, 12–20.

Kumar Verma, V., Arora, G., Mishra, A., and Rai, P. (2018). Generalized zero-shot learning via synthesized examples, in *Proceedings of the IEEE Conference on Computer Vision and Pattern Recognition*, pp. 4281–4289.

Kurle, R., Cseke, B., Klushyn, A., van der Smagt, P., and Günnemann, S. (2020). Continual learning with Bayesian neural networks for non-stationary data, in *International Conference on Learning Representations*.

Lampert, C. H., Nickisch, H., and Harmeling, S. (2009). Learning to detect unseen object classes by between-class attribute transfer, in *2009 IEEE Conference on Computer Vision and Pattern Recognition*. IEEE, pp. 951–958.

Lampert, C. H., Nickisch, H., and Harmeling, S. (2013). Attribute-based classification for zero-shot visual object categorization, *IEEE Transactions on Pattern Analysis and Machine Intelligence* **36**(3), 453–465.

Längkvist, M., Karlsson, L., and Loutfi, A. (2014). A review of unsupervised feature learning and deep learning for time-series modeling, *Pattern Recognition Letters*.

Larochelle, H., Mandel, M., Pascanu, R., and Bengio, Y. (2012). Learning algorithms for the classification restricted Boltzmann machine, *Journal of Machine Learning Research* **13**, 643–669.

Lecun, Y., Bottou, L., Bengio, Y., and Haffner, P. (1998). Gradient based learning applied to document recognition, *Proceedings of the IEEE* **86**(11), 2278–2324.

Lei Ba, J., Swersky, K., Fidler, S., et al. (2015). Predicting deep zero-shot convolutional neural networks using textual descriptions, in *Proceedings of the IEEE International Conference on Computer Vision*, pp. 4247–4255.

Lesort, T., Caselles-Dupré, H., Garcia-Ortiz, M., Stoian, A., and Filliat, D. (2017). Generative models from the perspective of continual learning, in *Advances in Neural Information Processing Systems*.

Lesort, T., Lomonaco, V., Stoian, A., Maltoni, D., Filliat, D., and Díaz-Rodríguez, N. (2019). Continual learning for robotics: Definition, framework, learning strategies, opportunities and challenges, *Information Fusion*. doi: 10.1016/j.inffus.2019.12.004, https://hal.archives-ouvertes.fr/hal-02381343.

Lessmann, S., Baesens, B., Seow, H.-V., and Thomas, L. C. (2015). Benchmarking state-of-the-art classification algorithms for credit scoring: An update of research, *European Journal of Operational Research* **247**(1), 124–136.

Li, L., Qian, W., and Clarke, L. P. (1996). X-ray medical image processing using directional wavelet transform. *1996 IEEE International Conference on Acoustics, Speech, and Signal Processing Conference Proceedings* **4**, 2251–2254.

Li, Z. and Hoiem, D. (2016). Learning without forgetting, in B. Leibe, J. Matas, N. Sebe, & M. Welling (eds.), *Computer Vision — 14th European Conference, ECCV 2016, Proceedings* (pp. 614–629). (Lecture Notes in Computer Science (including subseries Lecture Notes in Artificial Intelligence and Lecture Notes in Bioinformatics); **9908** LNCS). Springer. https://doi.org/10.1007/978-3-319-46493-0_37.

Li, Z. and Hoiem, D. (2018). Learning without forgetting, *IEEE Transactions on Pattern Analysis and Machine Intelligence* **40**(12), 2935–2947.

Li, H., Enshaeifar, S., Ganz, F., and Barnaghi, P. (2019a). Continual learning in deep neural network by using a kalman optimiser. *arXiv preprint* arXiv:1905.08119v3.

Li, K., Min, M. R., and Fu, Y. (2019b). Rethinking zero-shot learning: A conditional visual classification perspective, in *Proceedings of the IEEE/CVF International Conference on Computer Vision*, pp. 3583–3592.

Li, X., Zhou, Y., Wu, T., Socher, R., and Xiong, C. (2019c). Learn to grow: A continual structure learning framework for overcoming catastrophic forgetting. *CoRR* abs/1904.00310. arXiv:1904.00310, http://arxiv.org/abs/1904.00310.

Li, H., Dong, W., and Hu, B. G. (2020). Incremental concept learning via online generative memory recall, *IEEE Transactions on Neural Networks and Learning Systems*.

Lin, T.-Y., Maire, M., Belongie, S., Hays, J., Perona, P., Ramanan, D., Dollár, P., and Zitnick, C. L. (2014). Microsoft coco: Common objects in context, in *European Conference on Computer Vision*. Springer, pp. 740–755.

Lipton, Z. C., Berkowitz, J., and Elkon, C. (2015). A critical review of recurrent neural networks for sequence learning. *arXiv preprint* arXiv:1506.00019.

Liu, J. S. and Chen, R. (1998). Sequential Monte Carlo methods for dynamic systems, *Journal of the American Statistical Association* **93**(443), 1032–1044.

Lomonaco, V. (2019). Continual learning with deep architectures, Phd thesis, University of Bologna. doi: 10.6092/unibo/amsdottorato/9073, http://amsdottorato.unibo.it/9073/.

Lomonaco, V. and Maltoni, D. (2017). CORe50: A new dataset and benchmark for continuous object recognition, in Levine, S., Vanhoucke, V., and Goldberg, K. (eds.) *Proceedings of the 1st Annual Conference on Robot Learning*. Proceedings of Machine Learning Research (PMLR), Vol. 78, pp. 17–26. http://proceedings.mlr.press/v78/lomonaco17a.html.

Lomonaco, V., Desai, K., Culurciello, E., and Maltoni, D. (2019a). Continual reinforcement learning in 3d non-stationary environments. *arXiv preprint* arXiv:1905.10112.

Lomonaco, V., Maltoni, D., and Pellegrini, L. (2019b). Fine-grained continual learning, 1–14. *arXiv preprint* arXiv: 1907.03799. arXiv: 1907.03799, http://arxiv.org/abs/1907.03799.

Lomonaco, V., Maltoni, D., and Pellegrini, L. (2020). Rehearsal-free continual learning over small non-iid batches. in *CVPR Workshops*, Vol. 1, p. 3.

Lomonaco, V., Pellegrini, L., Cossu, A., Carta, A., Graffieti, G., Hayes, T. L., De Lange, M., Masana, M., Pomponi, J., Van de Ven, G. M., et al. (2021). Avalanche: An end-to-end library for continual learning, in *Proceedings of the IEEE/CVF Conference on Computer Vision and Pattern Recognition*, pp. 3600–3610.

Lomonaco, V., Pellegrini, L., Rodriguez, P., Caccia, M., She, Q., Chen, Y., Jodelet, Q., Wang, R., Mai, Z., Vazquez, D., et al. (2022). CVPR 2020 continual learning in computer vision competition: Approaches, results, current challenges and future directions, *Artificial Intelligence* **303**, 103635.

Loo, N., Swaroop, S., and Turner, R. E. (2021). Generalized variational continual learning, in *International Conference on Learning Representations*.

Lopez-Paz, D. and Ranzato, M.-A. (2017). Gradient episodic memory for continual learning, in Guyon, I., Luxburg, U. V., Bengio, S., Wallach, H., Fergus, R., Vishwanathan, S., and Garnett, R. (eds.) *Advances in Neural Information Processing Systems*, Vol. 30. Curran Associates, Inc., pp. 6467–6476. http://papers.nips.cc/paper/7225-gradient-episodic-memory-for-continual-learning.pdf.

Louizos, C., Ullrich, K., and Welling, M. (2017). Bayesian compression for deep learning, in *Advances in Neural Information Processing Systems*.

Mallya, A. and Lazebnik, S. (2018). Packnet: Adding multiple tasks to a single network by iterative pruning, in *Proceedings of the IEEE Conference on Computer Vision and Pattern Recognition*, pp. 7765–7773.

Mallya, A., Davis, D., and Lazebnik, S. (2018). Piggyback: Adapting a single network to multiple tasks by learning to mask weights, in *Proceedings of the European Conference on Computer Vision (ECCV)*, pp. 67–82.

Maltoni, D. and Lomonaco, V. (2019). Continuous learning in single-incremental-task scenarios, *Neural Networks* **116**, 56–73. https://doi.org/10.1016/j.neunet.2019.03.010, http://www.sciencedirect.com/science/article/pii/S0893608019300838.

Martin, C. H. and Mahoney, M. W. (2020). Heavy-tailed universality predicts trends in test accuracies for very large pre-trained deep neural networks, in *Proceedings of the 2020 SIAM International Conference on Data Mining*.

Masana, M., Liu, X., Twardowski, B., Menta, M., Bagdanov, A. D., and van de Weijer, J. (2020). Class-incremental learning: Survey and performance evaluation on image classification. *arXiv preprint* arXiv:2010.15277.

Maybeck, P. S. (1982). *Stochastic Models, Estimation, and Series on Mathematics in Science and Engineering*, Vol. 3. New York: Academic Press.

Merlin, G., Lomonaco, V., Cossu, A., Carta, A., and Bacciu, D. (2022). Practical recommendations for replay-based continual learning methods. *arXiv preprint* arXiv:2203.10317.

Mishra, A., Krishna Reddy, S., Mittal, A., and Murthy, H. A. (2018). A generative model for zero shot learning using conditional variational autoencoders, in *Proceedings of the IEEE Conference on Computer Vision and Pattern Recognition Workshops*, pp. 2188–2196.

Mitchell, T., Cohen, W., Hruschka, E., Talukdar, P., Yang, B., Betteridge, J., Carlson, A., Dalvi, B., Gardner, M., Kisiel, B., Krishnamurthy, J., Lao, N., Mazaitis, K., Mohamed, T., Nakashole, N., Platanios, E., Ritter, A., Samadi, M., Settles, B., Wang, R., Wijaya, D., Gupta, A., Chen, X., Saparov, A., Greaves, M., and Welling, J. (2018). Never-ending learning, *Communications of the ACM* **61**(5), 103–115.

Mnih, V., Kavukcuoglu, K., Silver, D., Graves, A., Antonoglou, I., Wierstra, D., and Riedmiller, M. (2013). Playing Atari with deep reinforcement learning. *arXiv preprint* arXiv:1312.5602.

Moreira, A. and Santos, M. Y. (2007). Concave hull: A k-nearest neighbours approach for the computation of the region occupied by a set of points.

Morgado, P. and Vasconcelos, N. (2017). Semantically consistent regularization for zero-shot recognition, in *Proceedings of the IEEE Conference on Computer Vision and Pattern Recognition*, pp. 6060–6069.

Neyshabur, B., Li, Z., Bhojanapalli, S., LeCun, Y., and Srebro, N. (2018). Towards understanding the role of over-parametrization in generalization of neural networks. *arXiv preprint* arXiv:1805.12076.

Nguyen, C. V., Li, Y., Bui, T. D., and Turner, R. E. (2018). Variational continual learning, in *International Conference on Learning Representations (ICLR)*.

Nguyen, C. V., Achille, A., Lam, M., Hassner, T., Mahadevan, V., and Soatto, S. (2019). Toward understanding catastrophic forgetting in continual learning. *arXiv preprint* arXiv:1908.01091.

Osawa, K., Swaroop, S., Jain, A., Eschenhagen, R., Turner, R. E., Yokota, R., and Khan, M. E. (2019). Practical deep learning with Bayesian principles, in *Advances in Neural Information Processing Systems*.

Pan, P., Swaroop, S., Immer, A., Eschenhagen, R., Turner, R., and Khan, M. E. E. (2020). Continual deep learning by functional regularisation of memorable past, in *Advances in Neural Information Processing Systems*.

Parisi, G. I. and Lomonaco, V. (2020). Online continual learning on sequences, in Oneto, L., Navarin, N., Sperduti, A., Anguita, D. (eds.), *Recent Trends in Learning From Data. Studies in Computational Intelligence*, vol 896. Springer, Cham. https://doi.org/10.1007/978-3-030-43883-8_8.

Parisi, G. I., Kemker, R., Part, J. L., Kanan, C., and Wermter, S. (2019). Continual lifelong learning with neural networks: A review, *Neural Networks* **113**, 54–71.

Parisi, G. I., Tani, J., Weber, C., and Wermter, S. (2017). Lifelong learning of humans actions with deep neural network self-organization, *Neural Networks* **96**, 137–149.

Parisi, G. I., Tani, J., Weber, C., and Wermter, S. (2018). Lifelong learning of spatiotemporal representations with dual-memory recurrent self-organization, *Frontiers in Neurorobotics* **12**, 78. doi: 10.3389/fnbot.2018.00078, https://www.frontiersin.org/article/10.3389/fnbot.2018.00078.

Patterson, G. and Hays, J. (2012). Sun attribute database: Discovering, annotating, and recognizing scene attributes, in *2012 IEEE Conference on Computer Vision and Pattern Recognition*. IEEE, pp. 2751–2758.

Pellegrini, L., Graffieti, G., Lomonaco, V., and Maltoni, D. (2019). Latent replay for real-time continual learning. arXiv:1912.01100 [cs.LG].

Perez, E., Strub, F., De Vries, H., Dumoulin, V., and Courville, A. (2018). FiLM: Visual reasoning with a general conditioning layer, in *AAAI Conference on Artificial Intelligence*.

Poggio, T., Voinea, S., and Rosasco, L. (2011). Online learning, stability, and stochastic gradient descent. *arXiv preprint* arXiv:1105.4701v3.

Politou, E., Alepis, E., and Patsakis, C. (2018). Forgetting personal data and revoking consent under the GDPR: Challenges and proposed solutions. *Journal of Cybersecurity* **4**(1), tyy001.

Pomponi, J., Scardapane, S., Lomonaco, V., and Uncini, A. (2020). Efficient continual learning in neural networks with embedding regularization, *Neurocomputing* **397**, 139–148.

Rao, D., Visin, F., Rusu, A. A., Teh, Y. W., Pascanu, R., and Hadsell, R. (2019). Continual unsupervised representation learning. arXiv:1910.14481.

Ravaglia, L., Rusci, M., Capotondi, A., Conti, F., Pellegrini, L., Lomonaco, V., Maltoni, D., and Benini, L. (2020). Memory-latency-accuracy trade-offs for continual learning on a risc-v extreme-edge node, in *2020 IEEE Workshop on Signal Processing Systems (SiPS)*. IEEE, pp. 1–6.

Rebuffi, S.-A., Bilen, H., and Vedaldi, A. (2017a). Learning multiple visual domains with residual adapters, in *Advances in Neural Information Processing Systems (NIPS)*, pp. 506–516.

Rebuffi, S.-A., Kolesnikov, A., Sperl, G., and Lampert, C. H. (2017b). iCaRL: Incremental classifier and representation learning, in *2017 IEEE Conference on Computer Vision and Pattern Recognition (CVPR)*, Honolulu, HI, USA, pp. 5533–5542, doi: 10.1109/CVPR.2017.587.

Rezende, D. J., Mohamed, S., and Wierstra, D. (2014). Stochastic backpropagation and approximate inference in deep generative models, in *International Conference on Machine Learning*.

Ritter, H., Botev, A., and Barber, D. (2018). Online structured Laplace approximations for overcoming catastrophic forgetting, in *Advances in Neural Information Processing Systems*.

Robins, A. (1995). Catastrophic forgetting, rehearsal and pseudorehearsal, *Connection Science* **7**(2), 123–146.

Robins, A. (2004). Sequential learning in neural networks: A review and a discussion of pseudorehearsal based methods, *Journal of Intelligent Data Analysis* **8**(3), 301–322.

Romera-Paredes, B. and Torr, P. (2015). An embarrassingly simple approach to zero-shot learning, in *ICML*, pp. 2152–2161.

Rostami, M., Kolouri, S., McClelland, J., and Pilly, P. (2020). Generative continual concept learning, in *Proceedings of the 34th AAAI*.

Roux, N. L. and Bengio, Y. (2008). Representational power of restricted Boltzmann machines and deep belief networks, *Neural Computation* **20**(6), 1631–1649.

Rusu, A. A., Rabinowitz, N. C., Desjardins, G., Soyer, H., Kirkpatrick, J., Kavukcuoglu, K., Pascanu, R., and Hadsell, R. (2016). Progressive neural networks. *arXiv e-prints* arXiv:1606.04671 [cs.LG].

Sariyildiz, M. B. and Cinbis, R. G. (2019). Gradient matching generative networks for zero-shot learning, in *Proceedings of the IEEE Conference on Computer Vision and Pattern Recognition*, pp. 2168–2178.

Sato, M.-A. (2001). Online model selection based on the variational Bayes, *Neural Computation* **13**(7), 1649–1681.

Savitha, R., Rajaraman, K., Krishnaswamy, P., and Chandrasekhar, V. (2018). Online deep learning: Growing RBM on the fly. *arXiv preprint* arXiv: 1803:02043.

Savitha, R., Ambikapathi, A., and Rajaraman, K. (2020). Online RBM: Growing restricted Boltzmann machine on the fly for unsupervised representation, *Applied Soft Computing* **92**, 106278, https://doi.org/

10.1016/j.asoc.2020.106278, https://www.sciencedirect.com/science/article/pii/S1568494620302180.

Schonfeld, E., Ebrahimi, S., Sinha, S., Darrell, T., and Akata, Z. (2019). Generalized zero-and few-shot learning via aligned variational autoencoders, in *Proceedings of the IEEE Conference on Computer Vision and Pattern Recognition*, pp. 8247–8255.

Schwarz, J., Czarnecki, W., Luketina, J., Grabska-Barwinska, A., Teh, Y. W., Pascanu, R., and Hadsell, R. (2018). Progress & compress: A scalable framework for continual learning, in *Proceedings of the 35th International Conference on Machine Learning (ICML)*. PMLR, pp. 4528–4537. *arXiv preprint*. arXiv:1805.06370v2.

Serra, J., Suris, D., Miron, M., and Karatzoglou, A. (2018). Overcoming catastrophic forgetting with hard attention to the task, in *International Conference on Machine Learning*. PMLR, pp. 4548–4557.

Shen, Y., Zeng, X., Wang, Y., and Jin, H. (2018). User information augmented semantic frame parsing using progressive neural networks, in *INTERSPEECH*, pp. 3464–3468.

Shin, H., Lee, J. K., Kim, J., and Kim, J. (2017). Continual learning with deep generative replay, in *Advances in Neural Information Processing Systems (NIPS)*, pp. 2990–2999.

Singh, P. R., Gopalakrishnan, S., ZhongZheng, Q., Suganthan, P. N., Savitha, R., and Ambikapathi, A. (2021). Task-agnostic continual learning using base-child classifiers, in *2021 IEEE International Conference on Image Processing (ICIP)*. IEEE, pp. 794–798.

Skorokhodov, I. and Elhoseiny, M. (2021). Normalization matters in zero-shot learning, in *International Conference on Learning Representations*.

Smola, A. J., Vishwanathan, S., and Eskin, E. (2004). Laplace propagation, in *Advances in Neural Information Processing Systems*.

Socher, R., Ganjoo, M., Manning, C. D., and Ng, A. (2013). Zero-shot learning through cross-modal transfer, in *Advances in Neural Information Processing Systems*, pp. 935–943.

Sodhani, S., Chandar, S., and Bengio, Y. (2019). Toward training recurrent neural networks for lifelong learning, *Neural Computation* **32**, 1–34, doi: 10.1162/neco_a_01246.

Sohn, K., Lee, H., and Yan, X. (2015). Learning structured output representation using deep conditional generative models, in *Advances in Neural Information Processing Systems*, pp. 3483–3491.

Song, J., Shen, C., Yang, Y., Liu, Y., and Song, M. (2018). Transductive unbiased embedding for zero-shot learning, in *Proceedings of the IEEE Conference on Computer Vision and Pattern Recognition*, pp. 1024–1033.

Subramanian, K., Suresh, S., and Savitha, R. (2014). A metacognitive complex-valued interval type-2 fuzzy inference system, *IEEE Transactions on Neural Networks and Their Learning Systems* **25**(9), 1659–1672.

Suresh, S., Savitha, R., and Sundararajan, N. (2011). A sequential learning algorithm for complex-valued resource allocation network-CSRAN, *IEEE Transactions on Neural Networks* **22**(7), 1061–1072.

Swaroop, S., Nguyen, C. V., Bui, T. D., and Turner, R. E. (2018). Improving and understanding variational continual learning, in *Continual Learning Workshop @ NeurIPS*.

Szegedy, C., Liu, W., Jia, Y., Sermanet, P., Reed, S., Anguelov, D., Erhan, D., Vanhoucke, V., and Rabinovich, A. (2015). Going deeper with convolutions, in *Proceedings of the IEEE Conference on Computer Vision and Pattern Recognition*, pp. 1–9.

Szerlip, P. A., Morse, G., Pugh, J. K., and Stanley, K. O. (2015). Unsupervised feature learning through divergent discriminative feature accumulation, in *Proceedings of the 29th AAAI Conference on Artificial Intelligence (AAAI-2015)*, Menlo Park, CA. AAAI Press.

Tao, X., Hong, X., Chang, X., and Gong, Y. (2020). Bi-objective continual learning: Learning 'n'ew while consolidating 'known', in *Proceedings of the 34th AAAI*.

Tarvainen, A. and Valpola, H. (2017). Mean teachers are better role models: Weight-averaged consistency targets improve semi-supervised deep learning results, in *Advances in Neural Information Processing Systems*, pp. 1195–1204.

Terekhov, A. V., Montone, G., and O'Regan, J. K. (2015). *Knowledge Transfer in Deep Block-Modular Neural Networks*, pp. 268–279.

Thrun, S. (1996). Is learning the n-th thing any easier than learning the first? in *Advances in Neural Information Processing Systems (NIPS)*, pp. 640–646.

Titsias, M. K., Schwarz, J., de G. Matthews, A. G., Pascanu, R., and Teh, Y. W. (2020). Functional regularisation for continual learning with Gaussian processes, in *International Conference on Machine Learning*.

Tomczak, M. J. and Zie, M. B. (2015). Classification restricted Boltzmann machine for comprehensible credit scoring model, *Expert Systems with Applications* **42**(4), 1789–1796.

Trippe, B. and Turner, R. (2017). Overpruning in variational Bayesian neural networks, in *Advances in Approximate Bayesian Inference Workshop @ NIPS*.

Tseran, H., Khan, M. E., Harada, T., and Bui, T. D. (2018). Natural variational continual learning, in *Continual Learning Workshop @ NeurIPS*.

Turner, J. T., Page, A., Mohsenin, T., and Oates, T. (2014). Deep belief networks used on high resolution multichannel Electroencephalography data for seizure detection, in *AAAI Spring Symposium*.

Van de Ven, G. M. and Tolias, A. S. (2019). Three scenarios for continual learning. *arXiv preprint* arXiv:1904.07734.

Van de Ven, G. M., Siegelmann, H. T., and Tolias, A. S. (2020). Brain-inspired replay for continual learning with artificial neural networks, *Nature Communications* **11**(1), 1–14.

Van der Maaten, L. and Hinton, G. (2008). Visualizing data using *t*-SNE. *Journal of Machine Learning Research* **9**(11).

Verma, V. K., Liang, K. J., Mehta, N., Rai, P., and Carin, L. (2021a). Efficient feature transformations for discriminative and generative continual learning, in *IEEE/CVF Conference on Computer Vision and Pattern Recognition (CVPR)*, pp. 13865–13875.

Verma, V. K., Mishra, A., Pandey, A., Murthy, H. A., and Rai, P. (2021b). Towards zero-shot learning with fewer seen class examples, in *Proceedings of the IEEE/CVF Winter Conference on Applications of Computer Vision*, pp. 2241–2251.

Wachter, S. (2018). Normative challenges of identification in the internet of things: Privacy, profiling, discrimination, and the GDPR. *Computer Law & Security Review*, **34**(3), 436–449.

Wah, C., Branson, S., Welinder, P., Perona, P., and Belongie, S. (2011). The Caltech-UCSD Birds-200-2011 (CUB-200-2011) dataset. California Institute of Technology.

Wang, L., Zhang, M., Jia, Z., Li, Q., Bao, C., Ma, K., Zhu, J., and Zhong, Y. (2021). AFEC: Active forgetting of negative transfer in continual learning, in *Advances in Neural Information Processing Systems*.

Wei, K., Deng, C., and Yang, X. (2020). Lifelong zero-shot learning, in *Proceedings of the 29th International Joint Conference on Artificial Intelligence*, pp. 551–557.

Wong, B. (2017). Data Privacy Law in Singapore: The Personal Data Protection Act 2012 (2017). *International Data Privacy Law* **7**, 287. Available at SSRN: https://ssrn.com/abstract=3456681.

Wu, C., Herranz, L., Liu, X., Wang, Y., van de Weijer, J., and Raducanu, B. (2018). Memory replay gans: Learning to generate new categories without forgetting, in S. Bengio, H. Wallach, H. Larochelle, K. Grauman, N. Cesa-Bianchi, and R. Garnett (eds.), *Advances in Neural Information Processing Systems*, Vol. 31. Curran Associates, Inc., pp. 5962–5972.

Wu, Y., Chen, Y., Wang, L., Ye, Y., Liu, Z., Guo, Y., and Fu, Y. (2019). Large scale incremental learning, in *Proceedings of the IEEE/CVF Conference on Computer Vision and Pattern Recognition*, pp. 374–382.

Xian, Y., Akata, Z., Sharma, G., Nguyen, Q., Hein, M., and Schiele, B. (2016). Latent embeddings for zero-shot classification, in *Proceedings of the IEEE Conference on Computer Vision and Pattern Recognition*, pp. 69–77.

Xian, Y., Schiele, B., and Akata, Z. (2017). Zero-shot learning-the good, the bad and the ugly, in *Proceedings of the IEEE Conference on Computer Vision and Pattern Recognition*, pp. 4582–4591.

Xian, Y., Sharma, S., Schiele, B., and Akata, Z. (2019). F-VAEGAN-D2: A feature generating framework for any-shot learning, in *Proceedings of the IEEE Conference on Computer Vision and Pattern Recognition*, pp. 10275–10284.

Xu, J. and Zhu, Z. (2018). Reinforced continual learning, *Advances in Neural Information Processing Systems (NeurIPS)*, **31**, 907–916.

Xue, G., Dong, Q., Chen, C., Lu, Z., Mumford, J. A., and Poldrack, R. A. (2010). Greater neural pattern similarity across repetitions is associated with better memory, *Science* **330**(6000), 97–101.

Yang, Y. and Newsam, S. (2010). Bag-of-visual-words and spatial extensions for land-use classification, in *Proceedings of the 18th SIGSPATIAL International Conference on Advances in Geographic Information Systems*, pp. 270–279.

Yang, Y., Chen, B., and Liu, H. (2022). Bayesian compression for dynamically expandable networks, *Pattern Recognition* **122**, 108260.

Ye, F. and Bors, A. G. (2021). Lifelong infinite mixture model based on knowledge-driven Dirichlet process, in *Proceedings of the IEEE/CVF International Conference on Computer Vision*, pp. 10695–10704.

Yoon, J., Yang, E., Lee, J., and Hwang, S. J. (2018). Lifelong learning with dynamically expandable networks, in *International Conference on Learning Representations (ICLR)*.

Zenke, F., Poole, B., and Ganguli, S. (2017). Continual learning through synaptic intelligence, in Precup, D. and Teh, Y. W. (eds.) *Proceedings of the 34th International Conference on Machine Learning*. Proceedings of Machine Learning Research. PMLR, Vol. 70. International Convention Centre, Sydney, Australia, pp. 3987–3995. http://proceedings.mlr.press/v70/zenke17a.html.

Zeno, C., Golan, I., Hoffer, E., and Soudry, D. (2018). Task agnostic continual learning using online variational Bayes. arXiv:1803.10123.

Zeno, C., Golan, I., Hoffer, E., and Soudry, D. (2021). Task-agnostic continual learning using online variational Bayes with fixed-point updates, *Neural Computation* **33**(11), 3139–3177.

Zhang, L., Xiang, T., and Gong, S. (2017). Learning a deep embedding model for zero-shot learning, in *Proceedings of the IEEE Conference on Computer Vision and Pattern Recognition*, pp. 2021–2030.

Zhou, G., Sohn, K., and Lee, H. (2012). Online incremental feature learning with denoising autoencoders, in *Proceedings of the 15th International Conference on Artificial Intelligence and Statistics (AISTATS-12)*, Vol. 22, pp. 1453–1461.

Zhuang, F., Qi, Z., Duan, K., Xi, D., Zhu, Y., Zhu, H., Xiong, H., and He, Q. (2021). A comprehensive survey on transfer learning, *Proceedings of the IEEE* **109**(1), 43–76. doi: 10.1109/JPROC.2020.3004555.

© 2024 World Scientific Publishing Company
https://doi.org/10.1142/9789811286711_bmatter

Index

A

approximate distribution, 56
approximate posterior, 57–58, 60
architect, regularize and replay (ARR), 101–102, 105–106, 108, 110–121
architectural, 101–102, 105, 110
autoencoder, 165–167, 169

B

backpropagation, 60
backward transfer, 183
Bayes by Backprop, 52
Bayesian VAE, 60

C

Caltech101, 184
catastrophic forgetting, 3, 5, 10–12, 14, 17–18, 22–23, 52, 56, 124, 145
catastrophic intransigence, 52, 56
chest X-ray classification, 197
Cifar10, 168, 172, 174
continual denoising, 180
continual learning, 74–79, 82, 91, 93, 99, 101–102, 104, 111, 113, 116, 118, 120, 197–198
continual reconstruction, 180
continual ZSL, 73, 76
coreset, 53, 58–59, 190
coreset VCL, 58

D

decision support system, 197, 201
denoising, 189
discriminative models, 54, 59

E

elastic weight consolidation (EWC), 61, 197, 199, 207–208, 214
encoder network, 60

F

fashion MNIST, 174
FiLM layers, 53, 71
flashcards, 163–165, 168–171, 173–174
Fréchet distance, 170
Fréchet latent space distance, 170
function-space regularization, 53

G

Gaussian, 172
generative models, 54

237

I

generative replay, 73, 75–76, 79, 82, 84, 86

I

ImageNet, 172, 174
intransigence, 3

K

knowledge distillation, 127, 176

L

Laplace propagation, 61
latent space, 168
lifelong ZSL, 80
likelihood, 53
local reparameterization trick, 53

M

mean-field approximate posteriors, 59
MNIST, 172, 174
multi-head architectures, 56
multi-head setting, 4–5

N

natural-gradient variational inference, 53, 70
negative backward transfer, 3
new-instance learning, 195
non-stationary data, 3

O

online learning, 2, 6–7, 26, 29–31, 33, 35, 40–42, 44, 49
online marginal log-likelihood, 59
on-the-job learning, 6–7
OOD detection, 197–199, 201–202, 208, 210, 212–213, 215
open-world environment, 6

P

positive backward transfer, 3
posterior distribution, 54
prior distribution, 53
progressive learning, 10, 16, 18, 20–22
pruning effects, 57

R

reconstruction, 170, 175
recursive, 167
regularization, 102, 105–108, 115
reinforcement learning, 12, 15–16
reparameterization trick, 62
replay, 101–102, 104–106, 109–111, 113, 115–121, 177
restricted Boltzmann machine, 28

S

sequential fine-tuning, 185
shared network parameters, 55
single-head setting, 4
Stein gradient method, 53
Stein gradients, 71
synaptic intelligence, 61

T

task-agnostic inference, 128, 142, 161
task-aware lifelong learning, 55
task-free lifelong learning, 56
task identifier, 123, 154
task-incremental, 125
task incremental classification, 181
task-incremental learning, 191
task-specific parameters, 55
true posterior, 58

U

UC Merced, 186

V

variational autoencoder, 54, 60, 74, 76, 78, 81
variational continual learning, 52
variational inference, 52, 57
variational lower bound, 53, 60
variational parameters, 60

W

wave height prediction, 47

Z

zero-shot learning, 73–75, 77, 79

Milton Keynes UK
Ingram Content Group UK Ltd.
UKHW020707240424
441680UK00002B/14

9 789811 286704